SASTUN

SASTUN

My Apprenticeship
with a Maya Healer

ROSITA ARVIGO

with NADINE EPSTEIN *and Marilyn Yaquinto*

HarperSanFrancisco
A Division of HarperCollins*Publishers*

HarperSanFrancisco and the author, in association with the Rainforest Action Network, will facilitate the planting of two trees for
every one tree used in the manufacture of this book.

FIRST EDITION

Library of Congress Cataloging-in-Publication Data

Arvigo, Rosita.

Sastun : my apprenticeship with a Maya healer / by Rosita Arvigo,
with Nadine Epstein, and Marilyn Yaquinto.

p. cm.

Includes index.

ISBN 0–06–250255–7 (hard cover); $22.00. — ISBN
0–06–250259–X (pbk.)

1. Mayas—Ethnobotany. 2. Mayas—Medicine. 3. Traditional
medicine—Belize—Cayo district. 4. Ethnobotany—Belize—Cayo
district. 5. Healers—Belize—Cayo District. 6. Medicinal plants—
Belize—Cayo District. 7. Rain forest ecology—Belize—Cayo
District. I. Epstein, Nadine. II. Yaquinto, Marilyn. III. Title.

F1435.3.E74A79 1994

615.8'82'0972825—dc20 93–37439

CIP

94 95 96 97 98 99 ❖ RRD(H) 10 9 8 7 6 5 4 3 2 1

Dedicated to Don Elijio Panti
Greg Shropshire
Crystal Ray Arvigo
James Arvigo
Mick, Lucy, Piers, and Bryony Fleming

Compañeros de mi vida.

CONTENTS

ACKNOWLEDGMENTS

I would like to thank: Greg Shropshire for his unfailing love and dedication and for making the many days away from home possible; Crystal Ray Arvigo for her patience throughout and assistance with the Creole language; Mike Balick for his professional guidance and cherished friendship; Lucy, Mick, Bryony, and Piers Fleming for dinner parties, comforting talks, and family events over the years; Daniel Silva for believing in us, and Florencio Marin for his confidence in the traditional healers of Belize; Katie Valk for her love and laughter; Polo Romero for sharing so much of his knowledge and time; Hortence Robinson for being my soul sister and teacher, and her daughter Concha Vasquez for her companionship; Angel and Isabela Tzib for their tender love and care of Don Elijio; Michael Usher for his invaluable expertise; Don Eduardo Espat for his confidence in Terra Nova; and Mary Kemmerer for her support over the years.

Special appreciation must be given to the late Timothy Plowman, who first believed in us and who gave us our first taste of plant collecting.

Thanks to Art and Ethel Arvigo and to Frank Arvigo for helping me to become the woman I am today. Thanks to Jenny Eshoo for the life she gave me. Thanks to Mosina Jordan, Barbara Fernandez, George Like, Steve Szadek, Paul Bisek, Mellen and Mohamed Tanomaly,

Georgina Vernon, and Melissa Bevans of the United States Agency for International Development for their assistance and grants to conduct the Belize Ethnobotany Project with the New York Botanical Garden and the traditional healers of Belize; the New York Botanical Garden for their longtime support; the United Nations High Commission for Refugees for their funding of the village community health care workshops on traditional home remedies; Carolyn and Jerry Garcia of the Rex Foundation; Katy Moran of the Healing Forest Conservancy; Mickey Hart; Steve King and Lisa Conte of Shaman Pharmaceuticals; Lu Nicolait of the Belize Center for Environmental Studies; Dee Dee Runkle, Sean and Yvette Bailey, Rob and Jane Mackler of the United States Peace Corps; and the residents of La Gracia Village in Cayo District for their support.

Over the years, the Belize Ethnobotany Project had been assisted by several foundations. We are grateful to the Rockefeller Foundation, the Metropolitan Life Insurance Foundation, the Overbrook Foundation, the Edward John Noble Foundation, the John and Catherine MacArthur Foundation, and the Nathan Cummings Foundation and the National Cancer Institute.

Thanks also to Charlotte Gyllenhaal-Huft; and Dr. Norman Farnsworth of the University of Illinois Napralert Service for providing scientific data on medicinal plants.

Thanks to Liz Pecchia, Bob and Nettie Jones, Katy Stevens, Carol Becker, Amini Awe, Alexander Woods, Darlene Domel, Julie Chinook, and Emily Ostberg for their assistance.

Special appreciation to Marilyn Yaquinto for her work on the manuscript.

I wish to thank the government and the people of Belize for creating a nation in which traditional medicine and traditional healers can thrive and flourish. May it always be so.

Thanks to the members of the Belize Association of Traditional Healers for their support and for helping me to keep the traditions alive by sharing their knowledge, and for their sacrifice of their own interests for the welfare and well-being of others.

Thanks to my critics, who have helped me steer a straight course in uncharted seas.

Thanks to Nadine Epstein for sharing my life and making me a better writer, and to our agent, Regula Noetzli, and our enthusiastic editors, Kandace Hawkinson and Andrea Lewis.

To *maestro* Dr. Elijio Panti, *"el mero,"* for his faith in me, for not letting me get too serious, and for placing the pearls of an unbroken chain in my hands. Finally, I am grateful to those on the other side of the gossamer veil for allowing me to peek in and for being there when I need them the most.

R.A.

I would like to thank for their love, faith, and support: Adam Phillips, Michael Epstein, Marcy Epstein, Donald and Jeanne Epstein, David Phillips, Gloria and Mike Levitas, and Lisa Newman.

Also Lonni Moffet, David McCandlish, John O'Leary, Leo Katz, Ricky Donald, Pat Dahl, J. P. Ferrie, Tracey Bohn, Maria Monthiel, Marta from Chaa Creek, Lee Oestreicher, J. C. Brown, and Sharon O'Malley, all of whom were there when we needed them.

Thanks also to Diane and Arlen Chase; Steve Houston; the Mexican Embassy in Washington, D.C.; *Ms.* magazine, *Smithsonian* magazine, and *The Whole Earth Review* for assigning the articles that led me to Ix Chel Farm; and Charles Eisendrath, the Michigan Journalism Fellows program, and the Kellogg Foundation for giving me the time to pursue my interest in traditional medicine.

I would like to thank Rosita Arvigo for her friendship and for living her life so fully and true; our agent, Regula Noetzli, for her nurturing of the project from beginning to end; and our wonderful editors, Kandace Hawkinson and Andrea Lewis.

Special thanks to my parents, Seymour and Ruth Epstein, who introduced me to the world of the Maya, and to Samuel "Noah" Epstein Phillips, who was born at nearly the same time as this book.

N.D.E.

Sastun (pronounced sas-toon):

The Mayan *sas* means light, pure, unblemished, and
mirror, while *tun* is stone or age. Together the
words can mean Light of the Ages, Stone of the Ages,
and Stone of Light, all of which are names for a
cherished tool of divination and spiritual power used
by Maya H'mens since ancient times. Sastun
can also be spelled zaztun or sastoon.

As part of my work, I receive a great deal of correspondence from people around the world. Many of these letters contain suggestions for plants that should be investigated, or invitations to visit people who have an interest in the relationship between plants and people. Some letters "glow" (for want of a better word) and stand out. They express sensitivity, strength, commitment, even intrigue.

In April 1987, I received such a letter from Dr. Rosita Arvigo, introducing herself and inviting me to visit the farm that she and her husband, Greg, had carved out of the tropical forest in Belize. She wanted me to meet "an old Mayan bush doctor" who, she wrote, "has practiced his ancient system of medicine for fifty years." No one in his community was interested in carrying it on beyond his lifetime, believing that his work was "with the Devil." Rosita, a naprapathic physician, herbalist, and now his apprentice, wrote that it would be a tragedy if Don Elijio's knowledge was lost to humanity.

Her timing was perfect. A group of us at The New York Botanical Garden had just received a five-year contract from the National Cancer Institute (NCI) to collect plants in the tropics of this hemisphere for testing in the NCI AIDS and cancer screens, as part of the NCI Developmental Therapeutics Program. Intrigued by her letter, I decided to stop in Belize that summer on my way back from Honduras.

I will never forget the day that I stepped off the plane, into a place that was to become one of the great passions of my life. With the warmth and gentility that is natural to them, Rosita and Greg were there to meet me at the airport. We drove across the country to their farm in western Belize and talked late into the night about our goals and philosophies.

The next day, notebook and camera in hand, I waded across the Macal River with Rosita, cut through the forest, and turned right on the dirt road that leads to San Antonio. Perhaps two hours later, we walked out of the forest onto a hill overlooking the small village of San Antonio. We then went to Don Elijio's house, where he greeted Rosita with the warmth and tenderness reserved for a favorite daughter. The three of us sat and spoke for hours about ethnobotany, Don Elijio's work as a healer, Rosita's apprenticeship, and the Garden's work with The National Cancer Institute. I was awed by his wisdom and strength of character.

Later that day, I watched as Don Elijio treated patients who came to see him. It didn't take long for me to realize that he was a powerful healer. My instincts about this special man were confirmed when at one time during the day he turned to me and asked a question about the National Cancer Institute's screening program and their search for chemical components of plants that exhibited cytotoxicity against living cancer cells. "Why must you poison the body in order to heal it?" he wanted to know. "If I were looking through the forest for plants that the Gods have given us for the treatment of cancer, I would look for something that would fortify the body rather than weaken it." In his one brief question, this elderly man from a remote village in Central America, who had never read a book nor seen a television, had crystallized the core of an important issue in the debate between various branches of modern medicine as to how to deal with this terrible human affliction.

Patient by patient, he explained to Rosita the basis for his diagnosis and the rationale for his subsequent treatment. "You see how swollen this baby's stomach is, and how the rest of the body looks, and how his pulse feels?" he would ask. Rosita absorbed every detail that was being presented to her. I was extraordinarily impressed with the commitment that Rosita had made to her teacher and to Maya traditional healing.

Late that night, sleeping in a hammock in a corner of Don Elijio's small house, I began to wonder where this journey could or would lead. Having been to so many different places in my botanical work over the last two decades, I was looking to settle down in a place where a botanist could make a difference. A place where very little work was going on and the obvious was being ignored.

The next day, as Rosita and I walked along the dusty road leading back to Ix Chel Farm, we began to talk. How could we combine our efforts to save not only the traditions but the rainforest itself? "Why not utilize a portion of our NCI contract to collect Don Elijio's plants, and screen these for modern medicine?" I suggested. "At the same time, we can begin to document his teachings from a botanical and health perspective," Rosita continued. Before we crossed the river we had agreed to a collaborative program that would involve the teachings of Don Elijio's and perhaps of other healers.

Later that afternoon, standing on one of the many hills at Ix Chel Farm, I looked out over the rainforest, at the crystal-clear river, and at my newfound friends. As it was obvious that we shared many of the same goals, we agreed to build this small collaboration into something that could affect many others, and somehow serve humanity at large. Thus was born The Belize Ethnobotany Project, a decade-long survey of the relationship between plants and people in Belize.

The rainforest I could see from the hill is, to me, one of the most beautiful places on the planet. In general, the tropical rainforest is one of the most spectacularly diverse habitats we have, containing nearly two-thirds of all of the plant and animal species that exist. There are many types of tropical forests, each with its own degree of diversity. Over four hundred different species of trees have been noted on a single hectare of tropical forest along the Atlantic Coast of Brazil, while a hectare of temperate forest near my home in Westchester County, New York, might have only five or six different tree species.

This diversity has extraordinary potential for human use. As Don Elijio likes to say, "for every ailment or difficulty on earth, the Spirits have provided a cure—you just have to find it." Yet modern science has not yet taken his advice. Fewer than one-half of 1 percent of the planet's 250,000 species of higher plants have been exhaustively analyzed for their chemical composition and medicinal properties. From that one-half of 1 percent, some 25 percent of all our prescription pharmaceuticals have been discovered.

In addition to medicines, tropical forests provide us with sources of food, fuel, fiber, dyes, and construction material, as well as the basis for numerous industries. But many benefits—such as diversity—cannot always be analyzed by an economist's pen. Maintaining diversity itself is a crucial goal for the world today, because with the reduction in biological diversity comes a total imbalance of the global ecosystem, which will eventually lead to its degradation and collapse.

As an ethnobotanist I know that one of our most important goals is to establish the value of the forest in a way that can be understood by modern economists and policymakers, as well as small farmers. In previous times, ethnobotanists focused on the production of lists of useful plants, sometimes combining nutritional or chemical studies with their explorations. Today, ethnobotany involves obtaining as complete an understanding as possible of the relationship between people and plants, from as many disciplinary perspectives as possible. This means that, in addition to identifying the useful plants, we need to understand the exact nature of their uses, how such resources are managed by people, how they are marketed and otherwise consumed, how they reproduce in the wild, and what their levels of sustainable harvest might be, as well as what are their physical, nutritional, or medicinal properties. Ethnobotany has evolved into an interdisciplinary science focused on the plant-people relationship at many levels, from that in a small village in a remote tribal territory to that in an urban center.

Sastun is a story of an extraordinary relationship between two people from two different cultures who find a common language in their love of traditional healing and plants of the rainforest. Its pages contain many of the lessons that Don Elijio has taught Rosita and, through this work, the world. This heartwarming story is one that also shows how much modern science can learn from traditional knowledge. Since 1987, we have collected hundreds of plants through working with Don Elijio, and these are now housed at the Belize College of Agriculture, the Forestry

Department, The New York Botanical Garden, and the Smithsonian Institution. Each of the plants contains with it information on its location, Mayan name, and Don Elijio's uses for it. Such specimens will last indefinitely and will continue to teach those generations interested in learning long into the future. In addition, bulk samples from these collections that were submitted to the NCI for testing are now being analyzed.

The Belize Ethnobotany Project has also involved studies with over two dozen healers in Belize, from a broad variety of cultural backgrounds. Using perspectives from people with very different backgrounds in Western medicine, traditional medicine, ethnobotany, and pharmacology and nutrition, this work has developed as a model for contemporary ethnobotanical studies elsewhere in the world. The project has also led to tropical forest conservation. Through Rosita's efforts, in June 1993, Terra Nova Rainforest Reserve was established in the Cayo District of Belize as the world's first ethno-biomedical forest reserve.

While Don Elijio's work and the traditions of Belize are being studied, tens of thousands of other traditional healers now face the prospect of erosion of their medical systems and plant resources as a result of acculturation, deforestation, and habitat degradation. I hope that this magnificent book, with its beautiful narrative, will serve to inspire others to carry out this kind of work in other parts of the world, with the same level of intensity, respect, and humility as has been shown through Rosita's special relationship with Don Elijio.

MICHAEL J. BALICK
Philecology Curator of Economic Botany
Director, Institute of Economic Botany of The
New York Botanical Garden
Bronx, New York

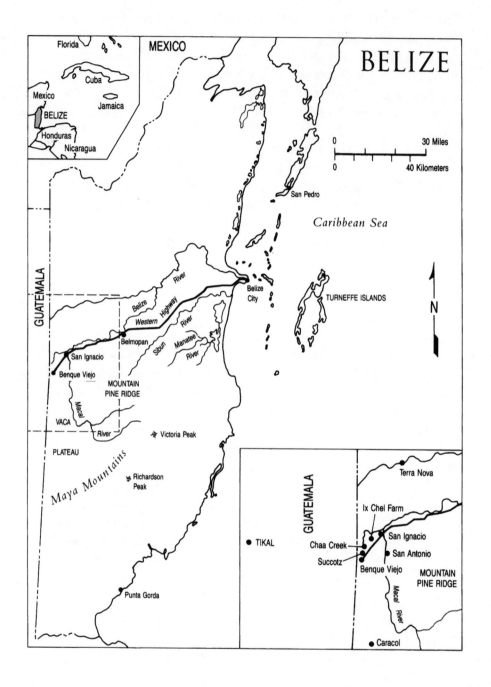

INTROCTION

Roses   Rosas   Nikte   Rosa chinensis

Red Roses have long been known to be useful in cases of infant
diarrhea, as a gargle for sore throat, and as an excellent skin wash
for rashes and sores. American Indian tribes dried the rose petals
and powdered them to use on infected sores and to blow into
the mouth to relieve sore throat. Central American women have
long relied on them as an effective means to staunch excessive
postpartum bleeding. Red Roses contain tannic acid, an
astringent commonly found in many plants.

One breezy, starlit, tropical night in Guerrero, Mexico, my life changed
forever. My deep and dreamless sleep was broken by an urgent knock
on my door, and I heard one of my neighbors, Doña Rita, calling my
name. Doña Rita, an arthritic Nahuatl woman of seventy-five, was on
her hands and knees at my doorstep. "My granddaughter is in labor,"
she explained breathlessly. "You must come to help deliver the baby."

I reeled back in horror and told her she couldn't possibly rely on me
for help, as I had never seen a baby born. I saw on her face the familiar
Mexican incredulity over North Americans' lack of real-life experience.
She firmly took my hand, and I numbly followed her down the steps to
her house. We worked through the night and delivered her granddaughter, Margarita, of a healthy baby boy. I was ecstatic, but Doña Rita
still looked worried. "Something's not right," she said. "There's too
much blood." She instructed me to go outside in the darkness with a
pine torch to fill a palm-woven bag with Roses and their leaves.

Dumbfounded, I did as I was told. She boiled the petals and leaves,
and when the mixture was cooled she spooned it gently into Margarita's
mouth and gave her the baby to suckle at the same time. In eight minutes the hemorrhaging had stopped.

This late-night event, dramatic yet so common in the non-Western world, changed my destiny. I wanted to know how those Roses saved Margarita's life. Doña Rita couldn't tell me much, only that she knew the red petals would stop the blood from pouring out. Her remedy remained an unsolved mystery until years later when I learned that the astringents in Roses and Rose leaves help stop bleeding.

That was 1973. In 1969, I had left my hometown of Chicago and a career in advertising to pursue my dream of living closer to the land. With my young son James, I moved first to San Francisco. Then, with friends, we left for Doña Rita's remote village in the Sierra Madre of Guerrero, where we farmed alongside the Nahuatl.

Since the closest government health clinic was an arduous fourteen-hour walk through steep hills and raging rivers, the Nahuatl relied on centuries-old formulas of herbal teas, baths, powders, and salves to meet their health needs. So when a family member became ill, the village elders—who possessed thousands of years of healing knowledge—were called in to administer household remedies that invariably worked.

Doña Rita and several other respected elders took me under their wings and taught me the names and uses of many medicinal plants. It dawned on me slowly but inexorably that the study of plants and their relationship to human illness would be my life's work. I had discovered that I had a gift for healing in my hands. I had no idea where this interest would eventually lead, but I had a sense that it would satisfy my yearning to be of service to God and humanity.

In 1976 I left Mexico, and the next year my daughter Crystal Ray was born in Belize—the former British Honduras—where I worked as a caretaker of an organic farm. When Crystal was two, we returned to Chicago, where I enrolled in the Chicago National College of Naprapathy, a three-year program that taught therapeutic body treatments that are an offshoot of chiropractic medicine.

There in cadaver class, I met Greg Shropshire, a handsome paramedic with beautiful, healing hands. We fell in love and were married shortly before graduation.

Greg and I decided to return to Belize. I missed the tropical climate, the year-round growing season, and the artist's palette of skin colors that paint the human landscape; the indigenous Maya and Spanish-speaking Central Americans, the Garifuna and Creole peoples with their roots in

Africa, the East Indians who had come as indentured slaves, the Lebanese who came as chicle-bosses, and the postcolonial Mennonites, Europeans, and Americans.

As alternative practitioners, we wanted to live in a country where medical freedom and traditional healing were still honored. Belize intrigued us because it had a thriving, highly respected tradition of *curanderos,* healers and herbalists, and there was no "medical practice law" that made natural healing a crime. We wanted to bring up Crystal, not yet six, in a healthy environment, swim in a pure river, eat home-grown vegetables, and live close to the natural rhythms of life in the untamed bush.

We decided to resettle in western Belize, buying thirty-five acres of uncleared jungle along the Macal River in the Cayo District, not far from the Guatemalan border. My dear friends Lucy and Mick Fleming, who had recently bought eighty-seven acres, called us when the land next to them went up for sale. We bought it sight unseen.

So in 1981, when James was entering college, Greg, Crystal, and I set out for our new tropical homestead. There we worked to clear and plant the land while maintaining a small practice as natural healers in the town of San Ignacio, six miles downriver.

Linden Flowers, Basswood   Flor de Tilo
Tilia americana

Linden flowers and leaves are a traditional home remedy in many
parts of the world for coughs, colds, and sore throats. Generations
of Central Americans have prized its effectiveness as
a mild, pleasant-tasting sedative tea for children
and the elderly and infirm.

It felt like a hot, sticky day in my hometown of Chicago. I could have
been catching a cool breeze on Lake Michigan, but I was on West Street
in San Ignacio of western Belize, and the only relief was the stingy
breeze from the Macal River. With a bag full of mangoes and papayas, I
stepped out of our clinic onto the veranda and looked out into the hot,
dusty bustle. I put down the bag and fanned myself, feeling too restless
to sit down. A wave of homesickness washed over me.

I loved San Ignacio, which is the largest town in western Belize and
only ten miles from the Guatemalan border, but today, the sweltering,
unforgiving heat; the loud, blaring *punta* rock music popular in Belize;
the row of shops along a narrow, dusty street; the mangy, threatening
dogs; and the pungent, powerful smells of everyday life seemed too foreign.

My husband, Greg, my daughter, Crystal, and I had been in Belize
for two years. Our dream farm, six miles away by dugout canoe, was becoming a nightmare. It was a never-ending struggle to keep the jungle
from encroaching on the two thatch huts we had built and called home.
Our effort to transform a muddy, burned-out clearing into a tropical

homestead was not the fulfilling experience we had imagined. We were seriously contemplating returning to Chicago. The indecision about whether to leave or hang on was frustrating.

I hated to quit, especially when I felt in my heart that we belonged in Belize. We were eking out a living from our natural healing practice, but even the herbs we'd brought from Chicago, which were our livelihood, were beginning to decompose from the unrelieved humidity. There was no way to replenish our supply. Clearly, our time was running out.

Shielding my eyes from the sun with a flattened palm I saw a diminutive old man sitting on our wait bench in the cool shadows. He had an air of learned patience. He seemed content to sit and watch the people pass by, peering at them with interest. His clothes were patched, badly stained, and threadbare, but freshly laundered.

"*Buenos días, señor,*" I said, stretching out my hand to greet him. He seemed startled and took a moment to compose himself before rising to greet me. His slight but sinewy frame barely reached my chin. Age had bent him slightly over at the waist, yet he had the bearing and presence of a much younger man.

He pushed back his sweat-stained yellow Pepsi cap, and I saw that his features were identical to those of the stone carvings at the ruins of the ancient Maya city-states such as nearby Caracol and Tikal. His face was a haunting rendition of the classic Maya profile: the long, hooked nose, the flat forehead, the drooping lower lip, and the upturned eyes.

Taking both my hands in his, he smiled warmly, looking into my eyes and speaking in a raspy voice, "*Mucho gusto. Mucho gusto.*" The feel of his tough, leathery hands conjured up the image of an experienced bushmaster slicing expertly through a tangle of wild vines. It was obvious the old man had spent decades in the jungle.

"Won't you come in and sit?" I suggested, wondering if he had come to see me as a patient.

"A friend left me here to wait for him while he runs an errand. Someone told me I might like to meet you," he explained in Spanish. "You are interested in the healing plants. Is this so?" he asked, carefully setting his stiff legs into a chair in the treatment room. He accepted my outstretched hand, and I helped him into the seat.

I explained that I was a natural healer, with a doctor's degree in naprapathy. "I use herbs, massage, and diet therapies in my healing practice."

He was straining his eyes to see what was in the quart-sized glass jars on the shelves behind us. "What are those?" he inquired.

"My herbs," I explained. "Would you like to see which ones I use in my work?" I pulled down one of the jars, preparing to launch into a primer about herbal medicine. But the old man stopped me and said, "First, let me introduce myself. My name is Elijio Panti of San Antonio Village not far from here, and I . . . "

As soon as I heard his name, I almost dropped the glass jar. I had been about to give a plant lesson to the best-known Maya medicine man in Central America.

Great and terrible stories circulated about this old Maya doctor-priest. Some spoke of near-miraculous healings, cured diseases, and numerous lives clutched from death's bony hand. Others claimed he was a lecherous old man, prone to molesting unsuspecting women, a drunk, a witch, a sorcerer, and a perpetrator of evil spells on innocent people.

I knew virtually nothing about local witchcraft beliefs except for gossip. Rumor had it that Elijio (pronounced Ay-leé-hee-o) Panti was from a family of black magicians. His father, it was said, was an *obeha* man, a practitioner of black magic who enchanted hundreds of women to be his lovers. I had heard that Panti also enchanted women, both for his own pleasure and for patients who paid for the service.

But as I looked down into the old man's gentle eyes, I found it hard to believe he was evil. I felt it more likely he was misunderstood, as healers often are. I too had been called a witch and had been accused of fanciful deeds.

Three of my patients had sworn that Panti had cured them of diseases no one else could even understand.

I asked him if he remembered a man who had suffered an awful wound, causing his leg to slowly rot away. "That man said you saved his life. He speaks very highly of your work, Don Elijio," I said, calling him by the title of respect Latin Americans use to address their distinguished elders.

He lifted his eyebrows as if trying to think back, but shrugged. He couldn't remember, he said. He had seen too many patients to be able to remember each one's story.

The old man was more interested in my herbs in their jars. He stared at them, peering into the shadows on the shelf. "What are these?" he asked as he pointed again.

"They're herbs from the North that we have shipped down here for our patients," I answered, delighted that a master was curious about my meager supply of herbs, disintegrating as they were. "You can see we're having a problem keeping them fresh and free of mold," I added.

"It's those glass jars," he explained. "They cause the moisture in the herbs to grow mold."

I was taken aback. His matter-of-fact comment had addressed one of our gravest problems in Belize: how to store dried herbs and protect them from the ubiquitous dampness. I wanted to thank him, but all I managed to say was, "Hmmm."

"I chop my herbs and dry them in the sun," he offered. "Then I place them in old cloth sacks inside the house away from the sun. That way they will last for months if you just dry them again outside once in a while. I've been doing this work for forty years, so I know a few things," he added, a devilish smile creeping across his wrinkled face.

He pointed with a scarred, crooked finger to a jar of Linden flowers. "Tell me, what do you use those for?"

"Oh, you may know that one as *Flor de Tilo,* the Mexican name. The buds are very good for nerves and sleeplessness. Easy to take, too. It has such a pleasant flavor." I was trying hard not to stare at him, but there was something oddly irresistible about the man.

He raised both hands in the air and yelped, "*Flor de Tilo! Mamasita,* I have not seen or heard of that blessed tree for many, many years. My dear, deceased wife came from Yucatán, Mexico, and her family often spoke of how much they missed that herb. It was gentle but sure, they said. And this one?" he asked, pointing at Buckthorn bark. I explained that I used this herb for stomachaches and constipation. He studied me as we chatted. "Your speech is Mexican. Are you a *Mexicana?*" he asked.

I told him that I had Italian and Assyrian parents but was born in the United States. I had learned my Spanish from living in Guerrero, Mexico, for seven years, studying plants with the Nahuatl Indian elders in a mountain village of the high Sierras.

"It's good to love God's medicines. They often cure when the doctor can't. People have to get help somewhere, so they come to look for *curanderos* like me and you."

I was flattered to hear him compare my work to his and thrilled by his interest and his curiosity about my herbs. I was succumbing fast to his charm.

He glanced over at my treatment table. I could tell he longed to stretch out and let me loosen the kinks and knots in his old frame. I was just about to invite him to experience a naprapathic treatment, when a noisy, dilapidated truck pulled up outside.

The old man's friend had come for him and was persistently honking the horn to get his attention. "I must go, *mamasita*," he said.

Just before he left, I gave him a bag of the dried *Flor de Tilo* leaves and flowers he had admired. "Ahhh, this will help me sleep," he said, with a wink.

Shaking his calloused, weathered hand, I knew I wanted to see this medicine man again. I heard myself ask him if I could visit him in his clinic so we could talk more about plants. "I have much to learn about the Belizean plants," I told him, "and perhaps I can help your stiff muscles."

"Yes, yes," he said, enthusiastically. "Come and visit me. I would enjoy such a talk, and my old muscles could use some beating and smashing." He let out a chuckle at the thought and shuffled out into the bright street.

He climbed into the truck and looked over his shoulder, waving his hand from side to side at the wrist like visiting royalty. The jeep spurted forward and lurched down the dusty, potholed street, turning the corner and nearly losing a wheel as it bounced out of a huge hole.

Despite the heat and the weighty decision about our farm's future, I felt light and happy after meeting the old Maya doctor. I imagined I felt a deep sense of integrity and simplicity emanating from him.

Still, I couldn't get rid of that voice in my mind's murkiest corner that he might be no more than a charlatan with a penchant for mischief. Worse, what if he misinterpreted my interest in him as a sexual overture? I'd learned from living in Mexico that many Latino men, of any age, mistake friendliness in North American women for open flirting.

My internal struggle was less about the man himself and more about having faith in my instincts. I had decided long ago to make life choices based on faith, not fear. My instincts had never failed me, despite having taken me on some pretty heady adventures during my forty-three years.

Putting the jars of rotting herbs back on the shelf, I wondered if Panti had an apprentice. Despite his vigor and bearing, he must, I thought, be in his eighties. When he died, would his profession disappear along with Central America's rainforests? Had any of his plants been studied by modern science? Was he all that was left of the extraordinary medical system of the ancient Maya, a last thread dangling off a once-glorious tapestry of healers who were revered, perhaps deified, in their society?

The thought that his knowledge of plants and medicine might fade into oblivion was heartbreaking. Would he want an apprentice? Would he teach me?

By the time I arrived back at the farm, I was preoccupied with Panti and rattled on to my husband Greg about the idea that had been buzzing around my head all afternoon.

"You don't mean that old witch doctor, do you?"

"Ah, let's not call him that," I scolded. "He seemed humble and spiritual to me." Then I told Greg about the glass jars. "Of course!" he almost shouted. "Why didn't I think of that? It's so obvious."

We finished the dinner dishes by the light of our kerosene lamps and put Crystal to bed on the wicker love seat that once belonged to my mother's living room set. An Indian bedspread separated her makeshift bedroom from the rest of the one-room hut.

As rosy dusk fell over the jungle outside our door, I suddenly realized that if, by chance, I were to work with Panti, we'd have to stay on the farm.

"How do you feel about staying in Belize?" I asked Greg, not sure how I felt about the big picture either.

"This is much harder than I ever thought it would be, Rose," said my exhausted husband. "I never seem to feel physically strong enough to keep up around here. I'm always tired. Every day I feel I have to scale a mountain just to be able to eat and bathe."

His face fell, jaw tight, when he spoke of the piece of jungle we had cleared away last month. It was already growing back and was several feet high. We were too broke to hire someone to help cut it down again. But it had to be cut down. Living too close to the jungle is dangerous: it brings debilitating dampness, bush animals, mosquitoes, and marauding creatures who abscond with food while you're sleeping.

"I think we were both naive about what it would be like to live here long-term," Greg said. There was despair in his voice. He got up to smash a scorpion meandering along our rustic, mahogany sink. With no running water, no electricity, and a road that was sometimes impassable, the litany of problems seemed endless. In order to get into San Ignacio, the nearest town, we had to canoe six miles down the river. Our life seemed like a perpetual and difficult camping trip.

"I'm exhausted too," I confessed. "I'm tired of worrying about money, and the dampness is like an ever-present enemy."

As a farmer, I knew the soil at our farm was poor. Despite the lushness of the jungle and the fact that the land had lain untouched for many years, the soil was impoverished, hardpan clay. It would take at least three years of hard work to build up its fertility with organic methods.

Greg and I reminded ourselves why we had come to Belize. We had known the going would be tough. We wouldn't have things like reading lights, hot showers, washing machines, or many of the other luxuries to which we were accustomed. Problems like the oppressive heat and rainy season would be endemic, just as city life had its crime and pollution. It was a matter of priorities, we concluded. And stamina.

"Ah, what the hell, we've been down before. We'll bounce back," said Greg, rubbing my shoulders affectionately.

"I'd like to go see Don Elijio, see what he's all about," I said, hoping Greg would share my dream about learning with the old bush doctor. "I'd like to ask him to teach me about the medicinal plants here. Maybe if we learned more about them, we wouldn't have to worry about the herbs from home going bad. I think we just have to keep on trying, love."

Greg nodded as he sat back on one of the portable massage tables we were using for beds. "Maybe Panti could give us a few pointers on this jungle living, eh?"

I chose not to pay much attention to Greg's comment. I was too busy thinking. "If he did agree to teach me, do you think we could afford it?" I thought out loud.

"That's going to be tough to do right now, Rose. I've got seven dollars in my pocket," Greg chuckled, as he pulled out the linings of his pockets to show off a few crumpled bills and a couple of stray nails.

We started laughing and called ourselves crazy people as we huddled by the open window and watched a silvery moon rise, sipping on Lemon Grass tea. The jungle, so black it was almost invisible, seemed like our friend again. As we listened to the chorus of river frogs and crickets, rising and falling in the night, we convinced ourselves that what we had dreamed was still possible.

We decided I would go to visit Don Elijio the following week.

CHAPTER TWO

Breadnut   Ramon   Chacox
Brosimum alicastrum

The fruits of the *Ramon* tree once provided a staple food to the
ancient Maya and Aztec Indians. The fruits are boiled and eaten
like young potatoes or ground into a cereal-like gruel and
sweetened. The cooked ground nuts combined with corn make an
excellent, nutritious tortilla and help to stretch the supply of corn.
The flavor of *Ramon* nuts is reminiscent of chestnuts. The leaves of
the tree are highly valued for female animals that have recently
given birth because they greatly increase and enrich the
milk supply. The diluted white, milky sap may be fed to newborn
infants as a milk substitute when mother's milk is not available.

The dry season had begun, and our farm surprised us by quickly transforming itself from a mud slide into a dust bowl. Until the grass we
planted patch by patch began to grow, we were going to suffer through
a cycle of mud drying into dust, dust churning into muck. This was taking a heavy toll on our fragile gardens, which either washed away or became parched in the blistering sun, leaving the soil cracked and split, as
if there had been an earthquake.

In the peach-colored dawn the family was enjoying a tropical breakfast of bread and fruits. It was one week after my first encounter with
Panti. This was Crystal's first day of school at Sacred Heart Primary
School in San Ignacio, and I was hoping it would be my first day of
school in San Antonio with Panti.

We climbed down the steep, slippery bank to the river. Mr. Thomas
Green, a tall, thin, dark-skinned Creole man in his seventies, was at the
helm of the twenty-one foot dory, which he'd fashioned himself from a
giant tubroos tree. I squeezed in among the children and schoolbags and
asked Mr. Thomas to drop me across the river. "I'm going to see the old
bush doctor, Elijio Panti," I announced, relishing the sound of my ad-
venture.

"You gwen go see di ole man?" he said excitedly, in the Creole-English that most Belizeans speak. "Dat good. Me like di bush med'sin. E good and di ole man know plenty," he added, while deftly maneuvering to the opposite bank. In Belizean fashion, he pointed with his lips, letting me know where the hidden trail began. It led up the hill connecting the bush to the main road to San Antonio. Waving good-bye to my daughter, I took the first step on my journey.

I was plagued with guilt about leaving Greg alone on the farm. He had the daunting task before him of hauling dead trees and logs into a pile, which would later be burned. With the onset of the winter rains, our homestead would become a snake den, with the creatures lurking under the dozens of piles of rotting brush we'd created. We still had so much left to clear, since the jungle was less than a dozen feet from where we slept at night.

He'd had such a small breakfast. I worried how he'd have the stamina to pile up tree trunks, and the mental image of his eating lunch alone was painful.

"Maybe I shouldn't go today, love," I had protested, but he had insisted that I go.

He had assured me he'd be able to cope, and our neighbor Mick Fleming was coming by to lend a hand. Mick and Lucy and their children were our only neighbors for miles. They were slowly and lovingly converting eighty-seven acres into an elegant jungle resort.

As I climbed up the bank the farm disappeared from view. Decades of treading bare feet had left an almost invisible sliver of soft, brown earth under the jungle growth. With one arm acting as a blind navigator I stepped into the thick brush. My machete tore through the vines. I took steady blows for a hundred feet until the workout tired me and my clothes were soaked with the sap-laden dew dripping off the plants.

I turned for a moment to gaze at the Macal River. It was steaming with a mist that swirled up from the surface like a Chinese dragon. Droplets of moisture ran off the trees and steadily plopped into the water, bursting into rainbow-colored rings of light that shimmered with the images above them. Hundreds of hidden *chachalaca* birds sang out in a chattering disharmony, piercing the morning quiet. This was a glorious moment for me. I felt welcomed by the forest and reveled in being surrounded by the abundant, luxuriant growth of plants and anticipating the key of knowledge that would unlock the secrets they held.

I remembered why we loved Belize so much. We were far from smog, the roar of traffic, and city grays. I stood above the bank, captivated by this magical glimpse of Mother Nature in her bedchamber, and sighed.

Red, biting ants crawled up my leg and rudely jolted me out of my dreamy state. I reached up inside my pant leg and brushed them off, then forged deeper into the jungle.

Early rays of sunlight streaked through the thick forest canopy, creating a shimmering jade glow. I startled a flock of rainbow-beaked toucans, which skittered away, reluctantly fleeing the tree where they were feeding on *Ramon* nuts. My eyes kept taking in new sights, although if I looked away from the path too long, needlelike thorns tore away at my flesh.

I followed the trail for nearly a mile, wandering through a wild profusion of waving palms, red shaggy-barked trees, aromatic vines, and swarms of iridescent insects and butterflies. Then a broad shaft of light ahead signaled the end of the cool, shady forest. Soon I stepped out onto the road and was quickly washed in white heat.

The road veered off in two directions from where I stood, and I nervously studied the tattered map. The Cayo District in which we lived was the westernmost part of the country and shared a long border with Guatemala. Belmopan, the capital of Belize, was in the eastern part of the district, while the town of San Ignacio was in the west. Scattered throughout were tiny villages like San Antonio, accessible only by rough roads and rivers. Our farm was six miles south—by river—from San Ignacio. San Antonio was five miles east of our farm, in the foothills of the Maya Mountains.

That sounded simple, but the roads were confusing. I started walking toward the right, wondering if I was lost. As far as I could see, I had five miles more of dusty road to go. And after a few minutes, sweat began running into my eyes, despite my straw hat, with its brim curled down to cut the penetrating glare. After two hours of hiking, I saw a whimsical, hand-painted wooden sign: "Welcome to San Antonio. Population 860." I walked up a steep, grassy knoll and saw the village unfold before me, with its red rooftops and rainbow-colored houses. They looked like a handful of hard candies spilled onto a lush, green carpet. Between the wooden and cement houses were carefully tended gardens with rows of sprouting vegetables. Brilliant bougainvillea in deep purple and warm yellow alamandas cascaded over wooden fences.

Maya people had lived in this valley surrounded by mountains for over a thousand years. Recently, archaeologists had excavated a field behind one of the village farms, finding an unusually large number of musical instruments in the more than three hundred homesites of the ancient Maya. The village, which archaeologists call Pacbitun, was ruled by the ancient city of Caracol, just thirty miles up the road.

I tried to imagine the ancient settlement, but modern San Antonio kept reemerging with its duet of barking dogs and blaring radios. Playful children ran up to me, singing out, "*Gringa, gringa,* gimme sweet." Their mothers scolded them in Mayan, and the children called back in a hybrid of Mayan, Spanish, and English, reminding me of a line from a Mexican revolutionary song: *niños mismo color de mi tierra.* Children the same color as my land.

Several women were turning peanuts they'd spread out on palm woven mats to dry in the sun. They waved as I walked by. I had seen no men, since it was September, harvest time for corn and beans in the fields just beyond the village.

Ahead on the road, three bronze, barefoot Maya women approached, balancing sacks of corn on their heads and babies in their arms. "*Buenos días.* Where, please, is the house of Elijio Panti?" I asked.

Their eyes darted playfully back and forth to each other, and they covered their mouths to mask their giggling. One pretty, almond-eyed woman pointed to the cluster of huts just inches from where I stood.

Panti's home and clinic reminded me of the old Chinese proverb: Sometimes, the greatest people in a village look like no more than a turtle in the mud. A dilapidated gray shack made of sticks and leaves leaned against a small, sturdy cement house with a zinc roof. Behind them stood a thatch hut with most of its walls torn away and a roof with gaping holes, open to both sun and rain.

A plump woman stood outside a general store just a stone's throw from Panti's front door. She eyed me closely before telling me he was out "*andando en el monte.*" Walking in the mountains. I followed her into her one-room store, crammed full of tins, chocolate, coffee, lard, cloth, brooms, and buckets of pickled pig tails. She opened the door to a rusty, gas-fired refrigerator, and I gratefully chose a cool tin of Guatemalan juice.

While she left to fetch her crying baby, I settled onto a stool in a shadowy corner. The woman returned and sat down near me to nurse her baby. She was Isabel, the wife of Angel, Don Elijio's grandson, she said, explaining that the old man was very busy, with patients trickling in all day long and sometimes into the night.

"Are you sick?" she asked, looking me over for telltale signs of disease. "You can wait in his kitchen. He should be back just before noon when the patients start arriving by transport."

The door to one of the thatch huts hung open, barely attached by a broken hinge. I stepped inside and the coolness of the interior surprised me. So did the state of disrepair. The walls displayed more stick than mud, and more than half the adobe was chipped away.

The room was no more than ten by ten feet. Three chopping blocks sat on the floor, surrounded by a dozen sacks of leaves, dried medicines, and corn. I could see no modern conveniences. It could have been A.D. 800 except for the nearby cement house with its zinc roof.

I had lived in huts like this one in Mexico. Panti's had a well-crafted clay hearth, where a brazen hen ruffled her feathers in a cloud of ashes. An enormous black pig grunted in the doorway and made a move to come inside and rifle through a basket of dried corncobs. "Cuchi, cuchi," I shouted and rushed at him. The pig stared dead blank into my eyes and shuffled away at his own pace.

"They respect no one and no place," said a man in Spanish, stepping into the hut with a woman and small child in tow. Once settled onto stools, we exchanged the usual friendly greetings. "My baby is sick," the woman said as she tenderly stroked the child's head. "She is four years old but as you can see looks no more than two."

The listless young girl looked up at me, igniting my healing instincts, with her arm dropped askew, pathetically out of her control. She drooled a bit and groaned, turning again to her mother for comfort. The child's eyes looked vacant and her breath was shallow. I could see her heart beating like a trapped bird under her paper-thin, lavender dress.

The girl had been vomiting and suffering from diarrhea for months, the mother said. They had taken her to clinics in Guatemala and in Mexico's Yucatán before running out of money.

"All the doctors say the same thing," said the mother. "'We see nothing on the machines, so there's nothing wrong with your child. She just needs better food and vitamins.'" However, a nurse in Guatemala had suggested they see the great healer, Don Elijio.

"Are there no *curanderos* in your own village near Merida?" I asked, surprised they had traveled so far to find a bush doctor.

"There used to be, but they're all dead and nobody learned," she said, letting out a doleful sigh.

It was true. Around the globe, traditional healers find it increasingly difficult to find apprentices to carry on their work. This is especially so in developing countries, where the need for traditional remedies still remains vital while the practitioners are dying out.

People don't want to appear "backward" by associating with these remaining healers. Many prefer the methods of Western doctors in white lab coats, who dole out expensive synthetic drugs—some culled from the very plants and herbs growing in their native soil. Christian missionaries have also made people feel ashamed of traditional healing by labeling it devil's work. Traditional healing has become so confused with black magic that many now fear working with spiritual forces, even when used for healing and service.

I craned my neck for some sign of Panti and spotted his small, sturdy frame trudging up the hill behind the huts. Through slits in the walls, I saw him inching up the hillside with a heavy sack strapped to his head in the ancient Maya style. I rose to help him, amazed at the weight of the sack as I lifted it off.

"*Buenos días,*" he said. "Just put that inside. First I must drink, then I will attend to you." I got a warm feeling seeing him again, but he barely noticed me as he glanced inside the dark hut to check how many patients were waiting. Within seconds, a smiling grandchild appeared with a brimming cup of *atole*. I recognized the warm, sweet brew made from ground corn kernels, traditional in Central America. With his drink, he disappeared into the cement house.

The transport truck from San Ignacio arrived at 11:30. There were no buses to villages off the main roads, so enterprising individuals bought trucks and ran on regular schedules, carrying up to thirty people, chickens, and farm supplies. Three people alighted, looking quite lost. The driver motioned in the direction of the kitchen, and soon seven of us were crowded into the tiny house. We watched the chickens pecking and scratching in the dirt floor near our feet.

As we waited for Don Elijio, a lively conversation sprang up. It was a warm and comfortable exchange, much more suited to a gathering of old friends than strangers waiting for a doctor. We chatted easily about the difficulties encountered on the road, the details of each person's ailment, and the hot weather.

When the patients spoke of Don Elijio, they respectfully called him *el viejito,* the old man, *numero uno,* or *el mero,* the authentic one.

Panti finally stepped inside and announced he was ready for patients. "I have been healing in this way, with my prayers, my roots, my vines and barks for forty years now," he told us, moving his arms about to punctuate each sentence. "I cure diabetes, high blood pressure, even cancer. I never went to school—cannot even sign my name—but up here, it's full." He tapped his forehead with a plant-stained fingertip.

The parents of the sick child got up, lifted their daughter, and followed him into the cement house. I peered into the room and saw the parents lay the girl down on a makeshift bed: an old door frame laid flat across two cement blocks. When they took the wrappings off her legs, I was shocked to see many infected, staphlike sores.

"The child has worms and *ciro* [gastritis] . . . and dirty blood," said Panti. "But *ciro* is her main problem. Oh, I know that old *cabron* [goat] Don Ciriaco well. He loves to fool doctors, but he can't trick me.

"There are three types of *ciro,*" he explained. "There is dry *ciro,* with constipation, red *ciro,* with bloody stools, and white *ciro,* with mucus in the stools."

First her digestion had gone bad and her blood become dirty. Then the worms had proliferated and the sores developed. "Dirty blood always has to come out through the skin," he explained.

"For me it is no mystery. She will be cured, *mamasita,* have no fear. God will help us all."

He ambled back into the kitchen/waiting room to fill a cloth sack with dried herbs, scooping them out of a large sack leaning against the hearth. He reached overhead to crush some dried plants hanging from the rafters, which I recognized as Epasote (Mexican Wormseed), and then he retreated to the cement house. Soon uproarious laughter came spilling out.

The same pattern ensued with each group of patients. "I've been healing this way for forty years," he'd boast dramatically, reminding me of an actor on stage. Then he'd take the patients to the cement house. He'd come back to the thatch hut for herbs. Then I'd hear laughter.

Hours later, he reappeared—showing no sign of fatigue—after having tended to all the patients. I was the only person left in the waiting room, and he now repeated his speech to me. "So what is your problem?" he asked.

"I'm not sick, Don Elijio. I'm Rosita. I've come to visit you. I am the woman you met last week in San Ignacio. Do you remember we talked about herbs and healing and I asked if I could visit you?"

He tapped his forehead. "Ah, it's these eyes, you know. Your face looks like a smoky mask to me. I am strong and stiff enough," he said, "to marry a fifteen-year-old." Except for his eyes, he lamented. "Now I wouldn't know if I was kissing a woman or a tree. Soon I won't be able to collect medicine for the patients."

"I would be happy to be your eyes in the forest," I offered. "I too am a healer and I need to learn about these plants."

"Oh, so you want to learn, child? It is good that you are interested in the plants, but I cannot teach you."

"I wouldn't be a bother to you, Don Elijio, I can see you're a busy man. I'll do whatever I can to be of service." He asked again if I was a *Mexicana,* and I repeated that I was an American from Chicago.

"It would do no good to teach a *gringa,*" he said sadly. "You must go home one day, it is only natural, and what I taught you would be lost up there. I am eighty-seven now and nobody here wants to learn. They come for healing, yes. But, where is the one who will open his heart to this hard work?"

Some of the villagers laughed at him, he said, calling him a *zampope,* or leaf-cutter ant, known for its ability to cut bits of leaves and carry them through the jungle to their underground lairs, rarely if ever stopping to rest.

"They say I have a pact with the devil. But they are wrong. I am lifting people up, not dragging them down. Never has a person walked in here and had to be carried out, but many were carried in and walked out after I healed them." He was arguing with unseen enemies.

There was another reason why he couldn't teach me, he said. His medicine came from the Maya Spirits. "Prayer is very important to my work, and our Spirits speak Mayan and you don't. Besides, my daughter, you have no sastun."

"What's a sastun?" I asked.

"It is the plaything of the Maya Spirits, and the blessed tool of the Maya healer," he declared, apparently expecting that would clear up any confusion and end my quest.

I didn't know what a sastun had to do with my learning about Belize's healing plants, but I decided to let the subject go . . . for now. I didn't want to pester him.

It was well past four and I had to head back to the trail before dark. Just as I began gathering up my things, several young Maya women entered, each carrying a baby. Glued to the hut door, I watched as Panti held each baby's chubby wrist and whispered in Mayan. He repeated this over the ankle, the other wrist, and the other ankle.

"Excuse me, Don Elijio, please," I said, observing proper protocol by requesting permission to return, "may I visit you again next week?"

"Of course, *mamasita*. I will be right here, neither more nor less than you see me right now."

As I walked out to the road, I could hear him and the young mothers speak the rich, mysterious sounds of the Mayan tongue. Then I heard the women squeal in delight, until the sound of my boots swishing in the grass drowned them out.

CHAPTER  THREE

Wormseed   Epasote   Chenopodium ambrosioides

A highly aromatic common weed found throughout the
Americas, used principally as a flavoring for beans, as a reliable
means of ridding intestinal parasites from children and adults, and
as a tea for flatulence. The entire plant boiled in water and drunk
throughout the day is a good cure for hangovers. For the bean
pot, add five leaves per quart of water when beans are nearly
cooked. For intestinal parasites in children, give one teaspoon of
the leaf juice each morning before breakfast for three consecutive
days. On the fourth morning give a teaspoon of castor oil.

————

I returned to Panti's clinic a month later. This time Greg rowed me
across the river and kissed me good-bye under a wild coconut palm.

It had started raining days ago and the path was slick and muddy,
defying all my feeble attempts to climb uphill. I slipped and fell twice,
caking thick, red mud onto my clothes and backpack. But I was determined. Soon the going got easier, and I studied the leaves on the fully
bloomed tropical trees and those in the thickets. Each plant seemed like
a stranger beckoning to me. I had always considered plants my friends
and was anxious to get acquainted with new friends in the Belizean
rainforest. Whether Panti helped me or not, I was determined to un-
ravel the riddles locked in the veins of a heart-shaped leaf or the fibers
of a clinging vine. I eyed a pale pink flower. Are you medicine? I wondered. I wished it could answer.

I was thrilled to see a familiar species—a thorny, flowering Wild
Poppy in brilliant yellow, growing on the edge of a sandy cliff near the
riverside. It was the first I'd seen in Belize. In Mexico it was called
Chicalote, and my neighbors had used it to treat insomnia and nervousness. My research had revealed that its active ingredient was papervine, a
proven, effective sedative.

I stopped near a mammey apple tree and took a gulp of refreshing peppermint water from my flask. I looked down to see a legion of leaf-cutter ants, which made me smile, thinking of Panti as the *zampope*. I watched them tirelessly mobilize the chips of green leaves—heavy loads for their tiny frames. I found that I couldn't wait to see the old man again.

I loped up the last hill to San Antonio. The distinctly pungent stench of pigs and rotting cornhusks greeted me as I walked the last quarter mile to Panti's compound.

He was at home, sitting on the kitchen floor, vigorously chopping medicine. He was chatting with the mother of the sick child, whom I had met the month before.

"*Buenos días,*" I said, as I came through the door. The woman introduced herself to me. Her name was Juanita, her daughter, María. "Yo soy Rosita," I said. They were camped out on the floor, and María was asleep on an old door, without a mattress or pillow.

In the Maya tradition, it's customary for healers to house their patients, since many are far from home and too poor to pay for room and board. By the time they had found Don Elijio, María's family had spent their savings on medicines, hospital stays, and taxicabs.

Panti didn't even glance up at me. He said a polite hello, then continued chopping. He didn't seem to notice or care that Juanita was at the moment straining to pull one of the wooden poles of the wall of his hut off its bracing to add to the fire. Already, more than half the wall was gone, leaving a large, gaping space. At this rate, the rest of the walls would burn up in a few weeks.

Juanita sensed my astonishment. "Too much rain this week," she explained meekly, as she stood surrounded by buckets and pots that had been called into service to catch the rain pouring in through the holes in the roof. "The wood is all wet and we must have a fire to warm María and cook our food."

"I built this house fifty years ago," Panti chimed in. "My wife and I lived here like two kittens on a pillow. Now the roof is rotten but those corner poles! They would outlive you and I both. They are of *Escoba,* a palm that doesn't even know how to rot. But this house is beyond saving. And when it is all gone, God will help me build another one."

When his wife had been alive, he explained, she had tended to patients, feeding them fresh beans, pumpkins, and homemade tortillas. She had always made sure the hut was snug and warm. Although he did his best, it could never be the same.

"It is too damp and muddy for me to venture out for wood—not with my rheumatism and these bad eyes—and the mother will not leave her child," he told me and shrugged.

I asked if there were any old logs and rotting fence posts around the village. Panti said there was lots of "widow's wood," so called because it is women without men who are forced to pick up scrap wood off the ground.

María awoke and sat up on her board bed. I was impressed by her improved appearance. Juanita began to feed her, and I watched as she took small gulps of warm corn cereal. The eyes that had been dull and vacant were now clear and shining. She held her body erect; there was no sign of listlessness. Even her hair, dull and brittle before, looked healthier.

"Ah, you look so much better, so pretty and much happier," I said. Her hand was frail and trusting in mine.

"There is no one like God and Don Elijio," said Juanita, smiling over at him while she rearranged her child's bedclothes.

I volunteered to go out and find firewood. "I'm already muddy, so it won't matter much if I hit the ground a few more times," I joked.

"That's because you are young and have so much blood," said Don Elijio. "When one is young nothing seems too hard. I am strong and stiff as a young man, and I would like to marry again—to a fifteen-year-old who will keep me warm at night and whisper secrets in my ear, and kiss, kiss, kiss," he said cheerfully while kissing the air around his elbow as if holding his invisible mate.

Juanita and I both giggled, which spurred him on.

"But these days I can't tell if I'm kissing a tree or a woman."

Juanita's eyes teared up as she laughed out loud.

"I need a blanket to keep me warm at night, a blanket of guts that turns when I turn and curls up around me." He wrapped his arms around himself in a symbolic hug. "A cloth blanket just falls off to the floor," he said, mimicking the unhappy sleeper.

Juanita finally threw up her hands and declared, "Don Elijio, you are shameless!"

"Yes, *mamasita*. Shameless and womanless. What is a man without a woman? Only half of nothing. It is much that a woman does for a man," he said, seriously enough to dampen our amusement.

"I need a woman to administer my house and to help with my patients," he said. "And my heart calls out for a woman, my body aches for a woman."

I left them chatting and went out to scrounge for wood scraps and came back loaded down with enough firewood to last the rest of the day. Then I made two more trips to a stash I'd piled under a tree near the edge of the village.

When I got back, Panti was tending to the child, who dangled her legs over the bed. He carefully washed out her open sores with a hot, green liquid, before taking out his tattered handkerchief and drying them. Then he reached under the bed for a musty glass jar with a rusted lid, and, tipping it on its side, he shook out a tiny bit of greenish black powder onto each sore. I noticed that some sores had healed while others were beginning to heal. The child winced as the powder fell, crying out to her mother, "*Me quema.*" It burns me.

With gentle assurance and ample confidence in his herbs, Panti spoke softly to the child. "Yes, yes, my heart. It burns, but it cures."

Piling the firewood under the hearth, I pulled out my machete from its leather scabbard and sharpened it with Panti's file. "May I help you chop medicine, Don Elijio?"

He peered over at me with his soft, weary eyes, asking, "Are you sick? Tell me first, because soon there will be many people arriving on the transport from town. It is better to tell me of your sickness now."

It was painfully obvious once again that he didn't remember me by voice or face. I gently reminded him that we had met twice. "Yo soy Rosita." His eyes crinkled at the edges and a smile creased his leathery face. He humbly apologized for his eyes. American doctors had visited the village the week before and informed him he had cataracts, leaving behind eye drops and sunglasses.

He pointed with his worn machete to the far corner toward another chopping block. "Those sacks hanging from the rafters—spread them out," he instructed. I did and then sat down on the dirt floor across from him and his hand-hewn wooden block. He pushed a pile of gnarled, brown vines in my direction and motioned for me to watch how he chopped. "Not too big. Not too small. Just so."

We worked in silence with only the sounds of the machetes on wood and the rustling of the vine tendrils. "What vine is this?" I asked, cutting through the quiet as gently as I could.

"This is Man Vine. This plant is for *ciro,* and its root is for men who can't."

I smiled at his delicate description then watched as he chopped. He wielded his machete without pause or worry, as if he were a master chef mincing vegetables with artful aplomb. Despite his failing eyesight, he was amazingly dexterous. At times, though, my heart leapt as his machete appeared to come dangerously close to hacking his already scarred fingers.

Juanita picked up María and arranged her on her lap as she sat in the room's only chair. The girl held a battered, naked, pink doll missing all its arms and legs. With one of her delicate fingers, she skillfully traced the lines around the doll's eyes and mouth, turning wounded plastic into a joyful toy. She bumped her thin, bare legs against the side of the chair, and some of the black powder Don Elijio had rubbed into her sores fell like coal dust onto her mother's already heavily stained dress.

"*Ciro* is something that jumps in your belly like a rabbit, but it is not a rabbit," Don Elijio continued without breaking the rhythm of his chop, chop, chop. "It is a very bad disease of the stomach."

Juanita interrupted, announcing that the *Epasote* herb, Wormseed, had run out. She had given María the last of it that morning. Despite the rain and mud, someone would need to gather more before dark.

I knew *Epasote* from my days in Mexico and volunteered to go hunt for it. Don Elijio was skeptical but handed me a quart-sized muslin sack and told me to try.

I found the plant about a quarter mile down the road, growing along a footpath near a creek besmirched by rusty cans and plastic bottles. I filled the bag with the fresh, aromatic leaves and returned to the hut.

Don Elijio was clearly surprised to see me return so soon. He seemed even more surprised when he inspected the contents and found

I had brought back the correct plant. Without a word he dropped the *Epasote* into a pot of boiling water waiting on the hearth.

The afternoon transport arrived, and soon four patients wandered into Panti's cramped kitchen. "Is this the house of Elijio Panti, the doctor?" asked a somber-looking man in Salvadorean Spanish.

"Elijio Panti?" shouted the medicine man, without missing a chopping beat. "That rogue! He's gone. You missed him. They chased him out of town long ago . . . said he was no more than a mad clown."

The family looked at each other in panic. They were just about to turn away when Panti stopped them and introduced himself. "I only jest like a mischievous boy. That is my way. I am the one you seek. Tell me, what is your problem?"

They were from the Valley of Peace, a settlement of Salvadorean refugees near Belmopan. The man's family had been sick for a long time, and nobody else had been able to help them. "We heard of you from a neighbor who sings your praises and prays that you will live many more years," said the Salvadorean, searching Panti's aging eyes with hope.

First, he pointed to his eighteen-year-old married daughter, standing next to him, and explained that she could not have children. The young girl blushed and dug her plastic shoe into the dirt floor.

"Humph!" said Don Elijio, waving his machete in the air. "Here I cure those that want and those that don't want. Nearly always it is the uterus that is in poor condition. This is a woman's center, her very being. Nothing can be right for her if her uterus is not in good condition. I massage, give teas, baths, and prayers. Then look out! She will have a baby for sure," he said, moving his hands over his belly to mimic the curve of pregnancy. He smiled over to the young woman, rocking an imaginary baby in his arms.

But the man still looked worried. He too had a problem, and it was more difficult to explain. "My luck has left me," he blurted out nervously.

It had started when he had been fired from his job as a watchman after a jealous co-worker told lies about him. Then, he and his wife, who had been happily married for twenty years, had begun to quarrel constantly. "I believe this is not natural," the man said. "Someone is doing this to us. It must be an enemy."

Panti nodded, then proclaimed with great forcefulness, "This is the work of the sastun! Come into the other house and we will pull out your luck to see if this is natural or not."

The family trailed off behind Panti. I ached to follow but got no invitation.

A minute later, I heard sonorous chanting. Panti was reciting in Mayan, and all I could understand was an occasional word in Spanish. Then I heard what sounded like the clanging of a clay object on wood. In the doorway I saw Panti bent over examining something in the light.

Panti held the man's right hand, then instructed him to open his palm. There was a small, translucent ball the size of a marble. Panti moved the man's hand back and forth, causing the marble to dance about on the flattened palm. Panti moved his face closer to the object, pushing up the glasses the doctors had given him and looking directly into it before pointing and exclaiming, "Yes. There it is. Do you see that black dot? That is your bad luck. That is your illness. Envy. Pure envy."

Panti smiled like a doting father at the man and continued his diagnosis. "Maybe you eat well, have a good job, have handsome cattle or beautiful, obedient children. Your neighbor begins to feel jealousy toward you. That makes you get sick in the head. You lose all your courage and drive. Don't worry, my friend, we can cure that easy. I understand all of this."

I had no idea what Panti was doing, but I had the feeling I was getting a glimpse of the heart of his work. It was clear that he was more than a man of plants. In some way, the cement house was as much a Maya temple as the ancient pyramids where shamans had healed mind, body, and soul with physical and spiritual means.

I thought, perhaps, that he was what I had heard called a H'men (Heh-mén). H'men translates as "one who knows." It was an honored title given to doctor-priests or priestesses of the ancient Maya civilization long before Cortéz and Bishop de Landa carved out the soul of the culture they found in the New World.

Juanita awoke me from my thoughts, saying, "The old man is very funny, isn't he? He laughs and jokes all the time—nothing is serious to him. I can see his loneliness and grief as a widower. Were I not a married woman, I would be tempted to stay with him. He is not poor and a very loving man. He would care for a woman well."

A short time later, the family filed out of the house and took seats on a roadside bench. Each person carried a bag of medicine. They looked cheerful and relieved, handing each other cookies and sodas. As they waited for a lift back to town, they seemed more like carefree tourists than the fretful patients they had been an hour before.

Panti came back to the hut and took his customary seat at the chopping block on the floor. His cheeks glowed.

I wanted to ask so many questions, but I didn't know how to start. I didn't want to pry or seem rude. Finally I said, "A lot of laughter in there, huh?"

He grinned and shrugged. "Oh, yes. Most people think too much, but get them to laugh and half of their trouble and sickness will go away."

We began to chop again, and soon the rhythm picked up where we had left off. Chop, thwack, chop, thwack against the worn blocks.

Within the hour, he climbed to his feet and announced he was off to tend to a new mother and her baby. I stayed in the hut chopping medicine until late afternoon, then swept out his yard and piled up the garbage left behind by his patients.

As I hiked home in the rain, I thought about what a funny old man he was. I was moved by his genuine compassion and empathy for people's fright and pain. Despite the gossip, I saw no "witch doctor." I saw a healer of the highest caliber and a talented clown. I saw a H'men.

Amaranth   Amaranto   Calalu   Amaranthus sp.

A favorite food throughout the Americas since ancient times,
when the toasted and sweetened seeds were molded with honey
into cakes offered to the Gods. Also known as "garden spinach,"
it can be prepared in any way that one would spinach. The
mature seeds make an excellent, protein-rich grain. A tablespoon
of the fresh leaf juice is given three times daily for anemia, as the
plant is rich in iron, calcium, and vitamins. The leaves
and branches are boiled and cooled to use as a
wash for wounds, sores, and rashes.

When I went to Don Elijio's clinic the next week, I found him alone. It
wasn't clear if he remembered me or not, but he welcomed, as usual, my
offer of help. He was pleased by my gift of a bag of European Chamomile, the last that we had brought from Chicago.

I brewed us each a cup of tea and Don Elijio began to talk. He told
me his story as we chopped a freshly harvested vine of pungent Contribo.

He was born in San Andreas, a small Maya village on a steep slope
on the Lake Petén Itzá in Petén, Guatemala. When he was only an infant, his father, Nicanor, killed a man in a drunken rage. At fifteen, his
father was already a known *hechisero,* one who practices black magic.
Authorities suspected he was responsible for a number of unexplained
deaths, and it was known that for the right price, he would use his
power to harm innocent people.

Rather than face justice, Nicanor fled. He brought with him his
wife, Gertrudes Co'oh. Nicanor had used his powers to enchant
Gertrudes when she was just fourteen. He had sent her sesame candy
contaminated by evil power that had rendered her a slave to his whims.

Gertrudes carried eleven month-old Elijio in her *reboso,* the ubiquitous shawl of Central American women. Sleeping by day and traveling

by night, they made their way through the jungle trails. After six days they reached the border of British Honduras, now Belize, and waded across the river under the cover of night. They joined thousands of refugees who came to Central America's only British colony to escape war and starvation. They resettled with Gertrudes's brother in Succotz, a Maya village alongside the Mopan River. There, Nicanor built a simple thatched house and planted corn to feed his family.

But soon Nicanor stopped caring for the fields or his wife and son. He began returning home late at night, often with other women. He'd pull Gertrudes out of bed by the hair, kick her around the floor, and force her to sleep under the stove, while he made love to another enchanted woman in their marriage bed.

Nicanor began charging large sums of money in Succotz to perform his evil spells. On occasion, he also cured a sick person using medicinal plants from the surrounding forests.

Young Elijio asked to learn about the plants but his father refused. "You have too much blood," Nicanor told him gruffly. "When you are older I may teach you, but not now."

At the age of nine, Elijio was put to work helping Uncle Isaac with his *milpa,* or field of corn, beans, and pumpkins. He was paid in corn, beans, and pumpkins, which kept his family fed. By the time he was thirteen, Elijio had secured his own piece of land from the village mayor, who felt sorry for him and Gertrudes.

The boy grew healthy and abundant crops. He was a good farmer because he had a natural love of plants and tended to his corn as if it were a personal friend. As is the old Maya custom, he showed his gratitude to his corn by saying prayers before he chopped down their stalks at harvest. Through plants he found peace and escaped the sadness of his violent home.

He wanted peace for his mother as well. Late one night when Elijio was fifteen, he lay awake, waiting for Nicanor to return home. His father kicked down the front door, crashing it against the wall. As Nicanor lunged for Gertrudes with ready fists, Elijio jumped out of bed and knocked his father to the floor. He forced a knee into Nicanor's chest and pressed the blunt side of a machete blade against his neck. Nicanor looked up in terror, twisting and groaning on the floor. Elijio shrieked, "I will kill you if you ever lay a cruel hand on my mother again! Sin or not, father, I will kill you!"

After that, Gertrudes was never beaten again.

Elijio labored to become an expert farmer. His beans were prized, and he traveled to another village in the mountains, San Antonio, to trade for leather, seeds, and chocolate.

One day he stopped to trade and talk with Damasio Tzib, a Mexican Maya from the Yucatán who had been one of the first settlers in San Antonio. Tzib's family had fled Mexico during the Caste Wars, the last Indian uprising against the Spanish. When they had arrived in British Honduras in 1906, Tzib told them, they had had a run-in with a naked, untamed clan of Maya still roaming the jungle around San Antonio. The Tzibs were drawing water from an old Maya well when the wild bushmen jumped out of the forest, brandishing bows and arrows and threatening to kill them. They spoke—remarkably enough—in the same Mayan dialect that the Tzibs spoke.

Elijio was fascinated by Tzib's story, but he couldn't keep his eyes off a young girl standing turning tortillas at the *comal*, a round clay disk fitted into hearths for cooking tortillas. She was Tzib's fourteen-year-old daughter, Gomercinda, known as Chinda. Elijio was smitten, relishing her beauty, her fleshy arms, mirthful eyes, and shiny, copper face. She wore the white cotton embroidered dress of Yucatán, and her long black hair was woven into braided cords and wrapped around her head. She smiled back at him with a coyness that signaled her approval.

As the young man walked back to Succotz that evening, he thought only of Chinda, muttering to himself in excitement, "She will be mine. She must be mine!"

But Tzib was reluctant. Nicanor's terrible reputation was widespread. After several months of courtship, Elijio's uncle and the mayor of Succotz convinced Tzib to trust him, although Chinda's mother warned she would reclaim her daughter if she was mistreated. He agreed, contrary to custom, to move to their village of San Antonio to protect Chinda from Nicanor's infamy.

There was never any need for her mother to worry, the old man told me. He felt only joy throughout his marriage to the woman he called the queen of his life. "We started out as children, but we lived together as lovebirds for sixty-five years," he told me. "The best part of my life has been loving a woman."

He always made a special effort to grow her favorite vegetables: tomatoes, sweet potatoes, Cilantro, and Amaranth, which she loved with hot tortillas and spicy salsa. She waited for him to return from the fields with freshly picked chilies, roasting them before grinding them into a sauce in her clay bowl.

He supported them by farming and trading. In the off-season, he worked in camps in the deep jungles of Mexico and Guatemala, where he collected Chicle sap, then used worldwide as the base of chewing gum. There he learned to drink too much, suffering bouts of alcoholism that would last throughout his life. Like all the men in his family before him, Panti fathered only one child, a girl named Emilia. For many years, Chinda and Emilia accompanied him to the Chicle camps, where Panti's Lebanese boss, Wahib Habet, paid them to cook and clean for the crew.

But when Emilia came to courting age, she and Chinda stayed behind in San Antonio. There Emilia married a man like her grandfather Nicanor. The man, Juan, was a brutal drunk who beat her regularly in front of the mournful eyes of their four small children and her horrified parents.

Many times Panti tried to step in and protect Emilia, but she told him not to interfere. He and Chinda felt helpless, listening to Emilia cry out in pain night after night.

When she was pregnant with her fifth child, they heard Juan beat her one night in a drunken rage. Panti broke his promise to Emilia. He lost his temper, grabbed an ax, and chopped down his daughter's door. Juan escaped out the back window, leaving Emilia sprawled out on the floor, drenched in her own blood. Her children sobbed and clung in terror to their grandfather's legs.

Emilia died soon after giving birth to Angel.

Panti scoured the bush for nearly two weeks, vowing to kill his daughter's murderer. "Sin or not, if I would have found that accursed man I would have chopped him to bits and felt no remorse."

Panti and his wife raised Emilia's children as their own. Soon afterward, Nicanor was killed by the father of a thirteen-year-old girl whom he had enchanted and seduced. Gertrudes buried Nicanor, then came to live with Panti in San Antonio. Chinda's mother, Teresa, also came to their home after Damasio Tzib died. Gradually, the household grew from "roots of despair into vines of happiness," Panti recalled.

Panti had always longed to be a healer. He had prayed to find the right teacher, one who would reveal the white art of healing, so he could forgo the insidious black magic his father took to the grave.

In 1935 a synthetic gum was developed that was cheaper than Chicle, and Panti's employers told him that this would be their last season. That season, the camp was in the Guatemala rainforest near the still-untouched ruins of the Maya city of Tikal. There Panti found his teacher, a mysterious Carib named Jerónimo Requeña.

One evening, after the moon had risen, the crewmen sat around the campfire drinking rum and telling boisterous stories. After a few drinks, Jerónimo bragged that he had the power to transform himself into a jaguar. The men grumbled and whispered to each other, goading him to prove his boasts and become a wild cat in front of many witnesses.

Jerónimo grinned, then picked up his shotgun. "Do not move from this circle. I will now walk into the forest. I will fire one shot and then you will see a jaguar climb up that ceiba tree behind you. I will pause and look down upon you, then I will disappear back into the jungle and return with the morning light."

With that, he slipped away from the campfire and tramped into the forest until he was out of sight. The men flinched when a gunshot rang out. Within minutes a massive, male jaguar crept into their view. The wild beast dug his claws into the bark of the ceiba tree behind them and sprang up to a sturdy branch high above their heads. A few men cried out, others crossed themselves, and many more ran for cover. The jaguar's eyes glowered, watching them scramble behind trees to hide themselves. He opened his cavernous mouth and roared ferociously, then crawled down the trunk and bounded back into the jungle.

Panti was too excited to sleep that night, and he kept an eye on Jerónimo's empty hammock. He knew from childhood stories that a true H'men could walk the night as a jaguar, totem of the H'men.

As sunlight cracked through the trees, Jerónimo appeared, with the strong odor of wild cat about him. Jerónimo slept for the rest of the day with his shotgun tucked under his arm.

A few weeks later, the rest of the crew left and Panti volunteered to stay behind with Jerónimo to guard the tools and equipment. They camped in a damp corner of one of Tikal's ancient temples, shrouded now by tree roots and formidable vines. "Over our heads were carvings in a mysterious design done by my ancestors," Panti said.

One night they roasted a monkey over a fire and got to talking. Panti was afraid of Jerónimo, and it took him a while to gather up enough courage to ask, "Tell me, *paisano* [countryman], do you know some things?"

Jerónimo looked over suspiciously and asked him to explain what he meant by "things." Panti said he wanted to learn about the healing plants. He explained about his father—a *curandero* who had wasted his gifts on black magic and refused to teach him.

He continued to press Jerónimo until the old Carib wearily answered, "Yes, yes, I know of these things, but I have no patience for healing people. They make me crazy. I don't mind patching up the *chicleros*, but I hide in these camps to get away from sick people."

The reflection of the fire flickered in the Carib's eyes as he turned and stared deeply into Panti's. "Do you have patience, boy?" he growled.

Without a moment's hesitation, Panti promised he was indeed patient and dependable. "Healing is what my interest is. Always I wanted to know, *papasito*. Please, will you teach me?"

Jerónimo turned his face back toward the yellow flames and remained silent for a moment more. He looked into the black of the bush and then spoke.

"There is no rest for the healer. Night and day they will come to your hut with their sad stories, their sickness. Their troubles are plenty. People do not understand the healer and often mistrust us. When we heal what the doctors cannot, the doctors call us *brujos*, witches, and whisper lies about us. They say we work with the devil. It is a lonely life, I warn you."

Panti kept nodding that he understood full well, and he continued to prod Jerónimo until the recalcitrant old man was won over by the young *chiclero*'s persistence. "Then I will teach you. Right now. Right here in the forest, where all the plants grow and the Spirits live."

There, deep in the thick, steamy jungles of Petén, his birthplace, Panti began his training as a *curandero*. He and Jerónimo searched out herbs, trees, vines, and roots. Panti had never learned to read or write, so he put everything Jerónimo said into his head. Each night by the campfire, they reviewed the day's lessons.

Jerónimo had him taste plants, make teas and powders out of them, and learn to recognize many of them while blindfolded. As Panti progressed, Jerónimo also taught him the Maya healing prayers that he had learned from his teacher and believed dated back to the ancient Maya H'men.

Jerónimo instructed Panti to have no relations with his wife whenever he was to cure a gravely ill patient or he would lose his power and she would become seriously ill. This he accepted with faith that Chinda would understand their new path together.

The weeks passed. The night before they were to leave, Jerónimo prepared for Panti's final blessing. He set out nine gourd bowls for a *Primicia* ceremony to introduce Panti to the Maya Spirits. Together, Panti and Jerónimo made corn *atole,* burned resin of the Copal tree, and said the *Primicia* chant.

As a final instruction, Jerónimo gave Panti the ancient and secret prayer that enabled him to stalk the night as a jaguar. But Panti didn't use it. He had no desire to become a cat. He was afraid that if he did, he'd be shot by a hunter.

Not long after, Jerónimo fell from a coconut tree and broke his neck. By the time Panti reached his side, Jerónimo was barely alive. He blessed Panti one last time and reminded him to always be kind and patient with sick people and to remember his *maestro* at future *Primicias.* The master then died in his student's arms, whispering, "I die happy because I have left it all to you. What you know will be my living memory."

When he returned to San Antonio, Panti began searching for the medicines in the nearby mountains and forests. "They were all there— by now my old friends." He gained experience in all manner of medical care, studying with midwives and Chinda's uncle Manuel Tzib, who had been a village *curandero* in Mexico. "I started to heal my family, then the villagers came, then people traveled from all around to reach me."

Only one thing was missing in his early practice in San Antonio. Jerónimo had told him that he would need a sastun in order to communicate with the Maya Spirits. "He who owns the sastun communes with the Maya Spirits as if they were close friends," Jerónimo had said. In the

Maya world only a gossamer veil separates physical from spiritual; by peering into the sastun a Maya H'men could determine the source of an illness or divine answers to questions.

Nine times a year for two years, Panti set up the *Primicia* altar in his cornfield and asked the Maya Spirits and God to send him a sastun to enable him to do their healing work better. One day his patience was rewarded. "I had just finished clearing up the nine gourd bowls from the altar, when I was overtaken by a great feeling of happiness. It made me skip and jump like a child all the way home."

When he reached Chinda, she was sitting under the lemon tree outside their kitchen. She put out her hand and said, "Look what I found on the floor today. A child must have left it behind. See, it's a marble."

In Chinda's hand, Panti saw the greenish, translucent stone that was his sastun.

CHAPTER  FIVE

Jackass Bitters   Tres Puntas   Kayabim
Neurolaena lobata

A common weed found growing throughout Central America,
much prized for its activity against parasites, including amoebas,
fungus, giardia, candida, intestinal parasites, and malaria. Either
fresh leaf juice or a boiled tea can be used for internal or external
purposes. Leaves and flowering tops of the plant contain an active
principle, sesquiterpene dialdehyde, an intensely bitter
substance found in many antimalarial plants.

A few months after my first visit, I arrived at Panti's doorstep at seven
o'clock in the morning, hoping to tag along and help him collect bush
medicines, but he'd already been *"andando en el monte"* for two hours.

The cement house was fairly new, built for Panti by his grandson
Angel after Chinda died so he would be protected at night. The two
thatch huts were only two feet apart: one a kitchen, where he chopped
and stored medicine; the other where he sometimes gave his patients
herbal baths and massages. Panti's good friend, Antonio Cuc, was at the
chopping block, cutting up a dark brown and yellow bark.

I sat down beside him. *"Buenos días, señor,"* I said. Don Antonio
seemed almost as old as Panti. He also had the classic square-jawed Maya
face, but his serious expression was in contrast to Don Elijio's twinkling
humor. His strong, calloused hands, criss-crossed with scars, wielded the
machete with practiced skill. He told me that he was Kekchi Maya and
that Panti was Mopan Maya.

There are an estimated four million Maya living in Central America
today, speaking twenty-five different dialects. Although the ancient
Maya had written glyphs, the dialects of the modern Maya are oral lan-
guages. Although he spoke Mayan, Don Antonio could no more read
an inscription on a Maya temple than I could.

"What are you chopping?"

"Billy Webb bark," he answered. "All this week I'll be clearing the high bush from my land. Next week I will burn."

"Why do you burn?" I asked.

"To prepare the land and make it ready for corn. My father did that, his father did that, and his grandfather did that. That is our way."

Belize still had over 50 percent of its old growth rainforest intact, more than any other country in Central or South America. But things were changing rapidly. The main culprits were land developers and commercial cattle and citrus companies, who were clearing thousands of acres at a time. Small farmers like Don Antonio only exacerbated the problem. They practiced "swidden" or "slash and burn" agriculture, a method used in Mesoamerica for more than five thousand years. Once efficient, the ritual burning had become a threat.

As I looked across the chopping block at Don Antonio's wizened countenance, I found it hard to think of him as a rainforest plunderer. Over the last months, I had seen that Don Antonio was one of Don Elijio's major herb gatherers. He routinely collected the medicinal plants on his land. Some he sold to Don Elijio, others he and his wife Doña Juana reserved for their own use. Like many elderly Maya, they were familiar with healing plants and respected their ability to cure. When they had raised their fifteen children, plants had been their only medicines.

As we talked, we saw Panti's small frame walk toward the huts. Don Antonio jumped up before I could to remove the heavy sack of bush medicine.

Don Elijio peered over at the patients that were sitting on the bench outside the door and remarked casually, "Ah, Rosita, you come again."

I was thrilled. At last he remembered me. I felt as though I had reached a mountaintop.

He and Don Antonio spoke in Mayan and I listened intently, but all I could understand were the occasional Spanish words mixed in. Despite the fame of the ancient Maya for numbers, mathematical concepts, and intricate calendars, the modern Maya use Mayan words for numbers one through five and Spanish for all others. Mayan speakers also use Spanish for the time, the days of the week, and the months.

Panti then went into his dark cement house to wait for his *atole* and to rest, leaving Don Antonio and me to keep each other company. I chopped with him. He told me that he used the Billy Webb bark to treat his wife's diabetes. She drank a tea made from the bitter bark for three months until she recovered. I listened excitedly, hungry for any tales about special plants that cure. I was sad when he left me alone at the block and went home for lunch. Then, to my delight, Panti popped his head in the hut and invited me to eat lunch with him. We sat at a table that was made from a crate, and his great-grandchildren brought his food. I pulled out my homemade granola and a thermos of apple juice.

As I chewed up my granola, he looked at me oddly. "What are you eating, child, mash?" I burst out laughing. Mash was the local term for chicken feed. I told him about granola and my vegetarian diet of fifteen years.

He smiled approvingly. Factory food was ruining people's diets, he scolded. People were being afflicted with what he called "modern food disease." "Junk" or *cuchinada* (pigged) food was at the root of most of his patients' ailments, which he noticed were worsening in recent decades. He said the intake of packaged foods—full of chemicals and preservatives—had made people more vulnerable to high blood pressure, heart disease, arthritis, diabetes, and cancer.

"For 'modern food disease,'" he said, "I give Balsam bark tea to cleanse the kidney and the liver, and many of these problems go away."

He also found grave harm in frozen popsicles, known locally as *ideals*.

"Since people starting sucking on those horrible things, they started with this *ciro*," he said. Only since the advent of refrigeration had people been able to drink cold drinks. "Too much cold makes the stomach cramp. After a while it stays in a knot, and one bite of food fills it up. Then when I massage the stomach, it has a giant pulse, it feels like a rabbit, but it is only *ciro*. If you take *ciro* to a doctor, he will shout, 'Hernia, hernia! Get the knife, we must operate!' But what can they take out, when it is just pure wind?"

I said I thought it was a shame that medicinal plants such as Man Vine that he used to treat *ciro* were being forsaken. I found this especially sad since modern medicine had found no better way to treat gastritis.

As I spoke, Panti chewed with toothless gums on the ancient regional diet of corn tortillas, beans, and hot chocolate. He said he abhorred the

Belizean favorite: rice and beans. And he didn't eat much of the other staples his neighbors favored, such as lard and pig tails.

Until very recently, most villagers had backyard gardens where they grew Chaya, Chayote, Cilantro, and some wild greens including Amaranth. Like Chinda had, they used to make salsa from fresh tomatoes, chilies, and garlic, and they drank a refreshing liquid made from orange leaves, Lemon Grass, and Allspice berries, providing them with vitamin C and minerals. Now they preferred to drink Coca-Cola.

"Let your food be your medicine and your medicine be your food," I quipped, quoting Hippocrates, the father of modern medicine. As a healer, I told him I encouraged the use of foods called "pot" herbs— plants we eat that also cure us, such as Rosemary, Amaranth, and Basil, which have medicinal power in their organic states.

I never wanted our conversation to end. We both looked out the window when we heard the afternoon transport rumble in, with Angel behind the wheel. His wife and several of their nine children were crowded next to him in the cab. Upwards of twenty people sat on wooden boards in the back of the canvas-covered truck amid farm supplies, chickens, and buckets.

That afternoon I was called into the cement house on several occasions to translate for patients who spoke only English. It was the first time I had ever been invited to help as Panti worked with his patients.

An eighteen-year-old San Ignacio woman, four months pregnant and paralyzed from the waist down, was one of the first patients. Two years before, she had had a very difficult delivery, and after the birth she had lost the use of her lower limbs. Panti said prayers into her wrists and forehead, prescribed herbal steam baths, and gave her a mixture of fresh green leaves.

She had tears in her eyes. "Have faith, my daughter, because God will help us all, if we but ask. Faith. That is what cures. I chop the medicine, I look for it in the jungle, I make the fires, roast the herbs, and apply them, but it is faith that heals."

Later in the afternoon, the young San Antonio policeman wandered in asking if Panti could do something for his chronic migraine headaches.

"Yes, there's a cure, but are you brave?" Panti asked. The policeman shifted uncomfortably and said, "Uhhh, yes, why do you ask?"

"Because the best thing for this is *pinchar,* he announced as he reached under his table, pulled out a dusty glass jar, and emptied a four-inch, bone-colored stingray spine into his palm. Recently archaeologists had uncovered a stingray spine in the tomb of a H'men. As Don Elijio washed the spine and the policeman's forehead with alcohol, I wondered if the archaeologists had ever seen a stingray spine in action.

Don Elijio stood up tall and confidently, pinched the flesh of the man's forehead between his fingers, and quickly pierced the flesh three times in three places, using the sharp end of the spine.

I heard "pop, pop, pop," as the stingray punctured the taut skin. The policeman grimaced but held his seat stoically.

Then Panti forced the man's head forward while streams of dark, frothy blood fell to the floor.

A foul odor filled the room.

"There you see your sickness on the floor," Panti said, pointing with his finger. "I've taken out the congested old blood blocked in your head, causing headaches. Blood should run like a river—clear, clean, and free."

The policeman looked relieved. "My pain is gone," he announced as he left. "I feel like a weight has been lifted."

After the patients left, Panti gave me an assignment: to toast a bushel of green leaves he had harvested that morning to treat a patient's skin sores.

"This is *Tres Puntas,* child. A very blessed plant. Keep the fire burning low under the *comal* and keep turning the leaves until they are toasted dry. I can't do it myself because I get burned, these eyes, you know."

After a few minutes of working with *Tres Puntas* I too could hardly see. Once the leaves began to heat up, they emitted a stinging smoke that burned my eyes and made my throat itch and my nose run.

Panti laughed and put another pile of fresh leaves on the *comal.* I moved the leaves about with a tree branch, trying to keep out of the shifting smoke, but finally gave up when he complained that I wasn't turning them often enough. In order to do a good job, I had to keep my face right in the line of acrid steam until the leaves were parched dry.

It was an especially torrid, tropical afternoon, and I was close to passing out with the added heat from the fire. But I didn't want to fail at my new task, so I blew on my upper lip and tried to cool my face; I had no hands free to wipe the sweat dripping down my neck.

He sat on a low stool in the breezy doorway, passing the crispy, almost burned leaves through a sieve into a gourd bowl. I watched as the leaves turned to blackish green powder, which he gathered into an old Tasters Choice jar with the label nearly rubbed off.

After I had finished, I went back to my post chopping the Billy Webb bark. Panti looked over at me with a proud smile. "I like the way you handle that machete, Rosita, but let me sharpen it for you. It will serve you better." We established a ritual. Every hour or so during our afternoon chopping sessions, he would stop his work to skillfully sharpen my blade with his file. It was a small bit of gallantry, his way of doing something for me.

We told stories to each other as we chopped. He loved to talk about his Chinda. She had been so trusting, never questioning him about treating women patients alone in their bedroom. "Even if I had to look at their private parts, never once did she pull back the curtain and say, 'What goes on in here?' No. Never would I put an evil hand on a woman patient. It's a sin. And I do not sin."

He told me the sad tale of Chinda's death.

"They killed her," he snapped, with enough fresh anger to convince me this was a wound that would forever fester. "Chinda had a hernia that I could not cure. In spite of my plants and prayers, it would not go away. I had to take her to the hospital in Belize City and let a doctor care for her. Imagine me at the age of eighty-one, making my first trip to Belize City."

Accepting that Chinda's illness was beyond his abilities, he returned to his village to continue harvesting his Christmas beans, planning to collect her within a few days.

"But I was in my cornfield when suddenly my heart began to flutter and pound in my chest. I dropped my tools and went home. People were gathering around my doorstep, waiting to tell me that Chinda's death had just been announced on the radio. *Mamasita,* with those words, I fell to the earth on my knees and fainted."

When he got to the hospital, he found Chinda's doctor upset and angry. He explained to Panti that the operation had gone well and Chinda would have recuperated without complications within a few days. But his instructions to give only water for the first forty-eight hours and broth on the third day were disobeyed. Instead of spoon-feeding broth to Chinda a little bit at a time, a nurse left the broth and a plate of heavy food unattended on the tray. When a hungry Chinda woke up, she gulped down the soup and the meal and died a few hours later.

"She didn't have to die. I shouldn't have left her. Had I been there, she would be here at my side today. I would never have let her eat that plate of heavy food. I would have taken better care of my patient."

Panti choked on the memory, finding little consolation in knowing the doctor had fired the nurse responsible that same afternoon. "For three years I was like a crazy person. I drank until I fell down in a stupor and cried myself to sleep in the dust. I had no room in my heart for sick people. I just wanted to die."

His cousin from the village of San Andreas and her husband came to stay with him for those difficult years, caring for him and trying to console his heartache. Sick people looking for the *curandero* found him drunk, sitting in the mud, or snoring in his hammock. "I'll never know how I survived those years. I knew nothing could be done; God gave Chinda to me and then he took her away. She was beautiful and fat— beautiful to me on the day I married her and beautiful on the day she died."

Silence fell on the room and filled in the gaps between the noise of our tools. I felt a great wave of tenderness for the warm-hearted old man as I watched him straddle a wobbly stool in his wife's crumbling kitchen.

A young village mother and her baby slipped quietly into the hut. Panti quickly regained his composure. He spoke to the woman in Mayan, then muttered a chant, holding her baby's wrists and ankles. The woman's two other children had come inside and were on the floor playing an old Maya game with nine stones that reminded me of jacks.

After the mother sat down on a stool and put the baby to her breast, Panti continued. "Life without a good woman at my side is like food without salt, coffee without sugar."

The young mother told him she felt sorry for him. He quickly rejected her pity. "I am still strong as a young man and blood runs in my veins," he boasted, jerking and pulling his arms to his side, shaking them to exaggerate his muscles.

"But no one here wants me. I've tried in my own village with three women, but they shamed me when they laughed at my courtship. They said I was too old. Yes, I am old, but my money is not old!"

With that, we all began giggling. He looked at me with his mischievous smile, and I saw again how much he loved to make people laugh. It was a sweet dose of the only medicine he could manufacture to treat his own illnesses—old age and loneliness.

The young woman left and as she did she said, "*In ca tato.*" That was the fifth time I'd heard someone say that as they left. I knew it was Mayan. I asked Panti and he told me it meant, "I'm going now, old revered one."

I glanced at my watch. It was late, and I told him I would need to leave soon since I had promised Crystal and Greg I would be home early to get ready for a party.

"How about one of your good treatments before you go?" he asked, placing his hand on the small of his back. He climbed on the bed and stripped down to the muslin shorts that Chinda had made for him.

I rubbed his back and neck with Wintergreen oil until his tired, overworked muscles began to relax under my kneading fingers. Panti moaned, "*Que rico,*" how exquisite, as I stretched and pulled his flesh. He had been suffering from rheumatism since the days when he had lived outside in the *chiclero* camps.

"You can do this to me anytime," he chirped. "Are you coming back next week?"

"I'll be here neither more nor less than you see now," I joked, giving him back one of his favorite lines.

In the fading daylight, he stood by his hut door, waving good-bye with a broad smile on his face. His cheeks had a pink glow that they hadn't had earlier.

But as I tromped past the sign for San Antonio on my way home, I knew he was sitting alone in his slanted-back chair, looking at the hen roosting in the cold stove where Chinda had once made her fresh tortillas and cheerfully listened to his stories.

Corn   Maize   Im Che   Zea mays

A sacred food to all cultures of Mesoamerica since preconquest
times. The grain is prepared primarily as a flat cake or "tortilla"
cooked on a clay disk called a *comal* and is made into a variety of
dishes. Corn Silk Tea is an ancient remedy for ailments of the
urinary tract, such as bladder infections or kidney troubles. A
hot, thick corn cereal called *atole* is a popular drink; mixed
with orange leaf tea, it is a household remedy for hangovers.
The four colors of corn—white, red, yellow, and black—
are believed to reflect the races of people, signifying the
four corners of the universe.

After a year of visiting Panti's clinic once a week, I arrived early enough
one morning to catch him before he set out for the bush. Past the flurry
of parakeets escaping the Sour Orange Tree in his yard, I saw him standing in the doorway of his cement house, adjusting his old plastic flour
sack around his shoulders and bending over to pick up his hoe. He wore
little black plastic boots and old plant-stained homemade pants. He
muttered incoherently to himself as he readied for the day ahead.

He was surprised to see me, but I was crestfallen when he said, "I
have no time for you today, child. The season is late, my corn is past
harvest time, and I've had too many patients to get to my own work." I
had always wondered about those sacks of yellow, white, and red corn
filling up his storage hut, still in their husks. I couldn't imagine that such
abundant and healthy ears of Indian corn were the fruits of his own
labor—not at his age and with his patient load.

"I'll help you harvest your corn, Don Elijio," I volunteered.

He looked incredulously at me, and as if to humor my enthusiasm
asked what I could possibly know about harvesting corn.

"Come on, *tato,* old revered one, I'll show you," I rebutted with
conviction. After all those years in Mexico, I knew how to harvest corn
like a veteran field hand.

He shrugged and agreed to let me tag along.

An iridescent orange sun had just broken over the horizon, yet the village was already bustling with activity. Women carried heavy loads of soiled clothes on their heads to be washed in the creek, schoolchildren in blue uniforms scampered around the yards, chickens squawked, and customers mingled in his grandson's store.

We walked along an old logging road in silence. I was perfectly content to quietly traipse behind him, following his little black boots as they keenly sidestepped the rocks and divots complicating the careworn, dirt road. Peanut fields spread out as far as I could see, while corn sprouted on the other side. Sprinkled here and there were patches of pumpkins and beans.

He eyed the herds of cattle chewing on barren fields, only recently converted from jungle bush, then shook his arm in the air and shouted over his shoulder, "Nowadays one has to go farther and farther to find medicine because they're chopping away so much of the forest. The fools don't know they're destroying life-giving medicine."

Yellow-head parrots bounded out of the cornfields as we approached, parading over our heads and flaunting their flaxen crowns. Panti looked up at the commotion and cursed them as shameless scavengers who can devour a day's supply of corn needed for a hungry family of five.

We passed a field of ripening watermelons glistening with insecticides. "Look at that," he said scornfully. "They are fooling themselves. They've poisoned their own mother and sent her out to sell herself in the markets of the world. She will make them pay for this cruel treatment. Payment always comes."

I was startled by his outcry but grateful for his sentiments, which echoed my beliefs.

Without breaking step, he continued, "Why can't farmers of today understand? The soil is like a bank account. You can only take out what you put in. Who has a bank account anywhere where you can only withdraw without making deposits? No. No, one must feed the soil with pure, good nutrients if you expect to get that back. Feeding the earth with poisons only means you harvest poisoned food. Not for me, I would never touch one of those melons. They are no more than artificial flowers."

This was a passionate topic for Greg and me, and I told Panti about our endless toiling over our stubborn soil. I stressed our refusal to hasten its production with chemical additives. I explained that a major factor behind our exodus from the States was to pursue organic farming and to live in a chemical-free environment.

"I'm willing to wait three years if I have to for a few good heads of lettuce," I added, chuckling because I'd probably have to, given the impoverished soil on our farm.

He nodded in approval, then led me off the main road toward a thin line of small trees and thorny shrubs that hid his cornfield from the road. As I crawled past the green barrier, my eyes lifted to grasp the masterpiece of cultivation spreading out before me. The sheer size of his handsome *milpa* was amazing, with stalks appearing to climb one on top of another until they reached a zenith at the top of a steep hill, spilling over into a sea of pale gold tassels and silk.

He sensed my wonder and spoke about his *milpa* with paternal pride. Corn was very important to the Maya. It was both food and medicine as well as a symbol of rebirth in Maya religious beliefs. The farmer buried the seed into the earth. It then died and went into the underworld. Then after the rain God Chac came and prepared the earth, the maize was reborn.

We climbed to the top of the hill, and he handed me a palm woven basket and pointed. I knew what to do and asked for no further instructions. Walking down the rows of tall, drying corn I began to pull off the mature ears and toss them into the basket, which I slung over my forehead, Maya style.

He looked over at me with bright eyes and a surprised smile on his face. I knew he was impressed, and I felt like I'd conquered a cultural wall that had separated us.

We worked in silence for hours in the cool morning air, but I couldn't help noticing that Panti took a new interest in me. I would catch him studying me as I rounded a row of corn or dumped a full basket onto a pile under a rough shelter he had built to keep his harvest dry.

When the sun coated our arms and slowed our rhythmic picking, I invited him to sit down under a Gumbolimbo tree and share an orange. As we sucked on the juicy sections, stopping to wipe our overheated brows, we exchanged tidbits about the crops we liked to grow best.

Suddenly he turned to look at me, peering curiously into my eyes. "Are you married?" he asked. This was the most personal question he had addressed to me over the last year.

"Yes, Don Elijio, I am married with two children," I said, slightly taken aback. It seemed that my ability to pick corn had suddenly aroused his interest in me.

"Oooooooohh," he said after a few seconds. A look of rejection crossed his face, and before I could tell him how much I'd grown to care for him he climbed to his feet. He towered over me and grew into a colossal figure against the blinding midday sun. Light seemed to emanate from him and dance on the golden cornstalks. I had never seen Panti look so powerful. Or so tall. My heart began to pound.

He said in a clipped, reproachful voice, "Just what is it that you want to know, my daughter?"

Without hedging, I said, "Don Elijio, if you will accept me as an apprentice, I promise to work hard and learn well."

Startling me, he pointed his finger and nearly shouted, "Do you have patience? Do you promise to take care of my people after I am gone?"

In rapid succession, he asked me about my plans for the rest of my life. Throngs of questions washed across my mind as if a mental dam had suddenly cracked open. Could I tend to the legions of sick people that crawled, limped, and stumbled to his door? Would his patients accept me as his apprentice? Could I now, right here, without even asking my husband for his opinion, agree to take Don Elijio's place?

The maelstrom of doubts and worries swirling about my brain were silenced by my convictions and commitment to healing. I heard my voice answer firmly, "Yes!" to all his questions.

"Yes, *papasito,* I promise."

Unconvinced, he continued hammering away at me, warning of the hazards of a medicine man's life. He outlined a picture of daylong hunts in search of a vital but elusive plant. Then picking, hauling, chopping, drying, and grinding the precious healing flora. "This is a lonely road, my child. Do not agree too hastily. *Curanderos* are often not even trusted by the very people they heal. They fear us, envy us, and some hate us. The gossip never ends. When we heal people that couldn't find help elsewhere, they call us witches. Then at night when you drop your weary body in the hammock, you hear, 'knock, knock,' and there is the person who called you a witch holding an infant on the doorstep of death. There is no rest by day or by night."

I told him I understood what his life was like. I'd witnessed his obligations and burdens for months on end. I wasn't afraid of hard work. "I want to learn, Don Elijio. I ache to know the names of the plants and how to use them to heal. Everywhere I go there are plants calling to me, but I know nothing about them. Please, I want to learn. I too am a healer, and I need your help."

He waved his arm and, without a trace of lingering doubt, surrendered to me. "Then, it is agreed. I will teach you." He reminded me, however, that without a sastun, he could only teach me so much. To be able to communicate with the Maya Spirits who lived beyond the veil, I would need a sastun.

Without a ceremonial handshake or toast to our good fortune—mine for uncovering an authentic teacher and his for securing a serious apprentice—we simply grabbed our tools and continued harvesting. The day was heating up and we still had to collect "Xiv" on the way home, he said.

Reveling in the glory of the permission of a student to ask questions, I boldly inquired: "What is Xiv?"

"*Xiv* is the Mayan word for medicinal leaf. There won't be time to go into the mountains today, but we will be able to fill our sacks with leaves along the way home. You'll see. Little by little, step by step, day by day." *Poco a poco, paso a paso, día por día.*

As I returned to my side of the cornfield, I thought about our pact. I still felt overwhelmed by the promises I had made, but I knew there was no turning back. My heart had made a dear and spiritually rich commitment that the rest of me and mine would, I hoped, one day embrace.

Spanish Elder, Buttonwood   Cordoncillo
Piper amalago

A common medicinal plant of many varieties, highly respected for
its versatility as a traditional remedy in Maya healing. The leaves
and flowering tops are boiled as a tea and used to wash all manner
of skin ailments, to aid insomnia, nervousness, headaches, swelling,
pain, and coughs, and for the treatment of all children's disorders.
The root is applied to the gums to relieve toothache. The raw
exudate of the root heals cuts and prevents infection.
The plant is one of the Nine Xiv used by Don Elijio
for herbal baths.

With so much to learn about the medicinal plants of the Maya, I decided to stay with Don Elijio three days each week.

I slept in the cement house with him. My hammock was stretched
across the length of the waiting room, separated from his small room by
an embroidered, orange curtain, the last piece of handiwork Chinda accomplished before she died. In Spanish, its fitting inscription read: "I
will love you forever." The words swirled around two bluebirds on a
flowering branch.

The first morning I was there he tugged at my hammock strings.
"Wake up, child! No time to lose," he said in a raspy whisper.

It was a chilly winter dawn and I groaned. I hadn't slept well the
night before, unaccustomed as I was to sleeping in a hammock. On re-
flex, I yelped, "*Sí, maestro,*" before taking a deep breath and swinging
my legs onto the cold cement floor. He turned his back while I dressed
in yesterday's clothes and tied up the hammock where it was stored during the day. He washed his toothless mouth with water from a bucket
stored in the corner and gave me a cup of water to wash with.

Breakfast for Don Elijio was white sweet bread dipped in a cup of
instant coffee, mixed with three spoons each of powdered milk and

sugar. I poured myself some hot chocolate from my thermos and munched on cinnamon crackers.

"Eat quickly, child. *C'ox c'ax,*" he said in Mayan, which meant, "Let's go to the forest." I gulped down the last of my warm drink and started strapping on plastic flour sacks, machetes, picks, and water containers.

Once outside, looking into the rising sun, my sleepiness fell away. I have always loved the early morning hour. Here it was vibrant with bees, crickets, and other insects. The old logging road to the rainforest skirted the foothills beneath the Mountain Pine Ridge and wound through peanut fields and small plantations of banana and cassava. The hills beyond us were blanketed in a vista of fan palms and crests of flowering trees in radiant orange and yellow. The last houses, nestled in lush green land and painted in vivid colors, mimicked the bold shades of flowers and the bright, breasted birds.

Don Elijio stopped before a bush. "Xiv," he said as he pulled off its leaves and tenderly stuffed it into his sack. With each yank of a stem, he muttered under his breath.

I remembered I now had an invitation to ask questions, so I confidently asked, "What is this plant used for?"

Like a symphony playing in my ears, he answered. "This is *Anal.* You will remember it by its bunch of flowers at the top, which are first green then turn white and make red berries just before the next rain comes."

Xiv (pronounced shiv), he said, is the daily collection of at least nine medicinal leaves to be used that day for baths and teas. "I know ninety pairs of Xiv," he continued. "Pair by pair. Ninety males and ninety females. Each with a name. It's all up here in my head," adding his customary tap on his right temple.

I asked what he whispered while he snipped off the leaves.

"The *ensalmo.* You must learn many," he said, as if I knew what he was talking about. I was surprised that despite my fluent Spanish, I had never heard the word before. I thought it might be a prayer, but I thought better of asking.

Instead, I asked Panti why he did the *ensalmos.* He stopped walking and looked at me sideways, shaking his head in disbelief. "It's simple, my child, if you don't thank the Spirit of the plant before you take it from the earth, it will not heal the people. Many people say they gather what they see me gather, but it doesn't work for them. That's because they haven't remembered to say the *ensalmo*."

It seemed so natural and almost too basic not to have been a part of my thinking before. The plants are living things, and they give up their lives to help us heal. A prayer seemed like a small price to pay for such a gift, I mused.

He repeated the *ensalmo* for me ever so slowly: "I am the one who walks in the mountains seeking the medicine to heal the people. I give thanks to the Spirit of this plant, and I have faith with all my heart that this plant will heal the sicknesses of the people. God the Father, God the Son, and God the Holy Spirit. Amen."

"This one with the white string flower is *Cordoncillo*," he said. "There are many kinds." He crushed a leaf and held it to my nose. Then he had me taste the flower, and I was reminded of licorice. "*Cordoncillo* should always be added to the Xiv mixture and is strong medicine on its own." Later I learned the plant was known in English as Spanish Elder.

"And, Rosita, when you're gathering medicine, never say to yourself, 'I hope this works,' or 'Maybe this will work.' No, no, you must say with complete confidence in the plants and faith in God that these plants will heal. And they will, I promise you."

I told him not to worry, I was a person blessed with a great deal of faith.

"It is good to have faith. That is the most important lesson you will learn from me," he continued. "With faith, everything is possible. Believe this, for it is true."

As we traveled deeper and higher into the forest, the soil changed from thin and rocky to soft and rich. It was damp and dark and the aroma of decaying humus filled the air.

Panti's long cutlass bounced against his leg as he walked. He took the cutlass out of its leather case whenever he needed to cut away a vine or branch impeding our progress.

"Humph. This vine thinks I don't have a machete," he said, ever so gently slicing back only what was necessary to make our way through.

Every now and then on the trail, he touched the end of a toxic plant with the tip of his machete as a warning. He told me about Wild Chaya, which bears stinging white hairs that can easily penetrate through layers of clothing. When it touches the skin it forms blisters that burn for days.

He showed me the delicate vine with lovely white flowers called *Lindahermosa* ("pretty beautiful" in Spanish), which has thorns that cling with barbs, tearing out pieces of flesh as you pass by. Then he warned about the Cockspur or *Zubin* tree that signals its danger with its bark blanketed with fat, razor-sharp thorns that harbor biting ants.

We'd been steadily trudging uphill, and I was panting and stopping frequently to catch my breath. Don Elijio, with nearly fifty years on me, was hardly breathing above normal, as if this journey were a leisurely stroll through a city park. He rarely if ever stopped to rest. Like the *zampope,* the leaf-cutter ant, I thought, smiling, watching the back of his spindly legs in his little black boots.

I caught up with him in front of a tangled canopy of green vines and small waxy leaves. He put down his burdens and sliced through the thick lianas to get to the center where the root system was hidden. He motioned to me to come closer, and I caught a glimpse of an exposed bit of a gnarly, black root trunk, which he was scratching with his thumb. He smiled and hailed, "Aha! It is good luck to collect medicine with a woman companion. The Goddess of medicine, Ix Chel, has her subjects show themselves to the healer more readily. Smell this root, my daughter, and remember."

"Ix Chel, who is she?" I asked while scratching and sniffing the ebony root. I wasn't at all prepared for the foul stench.

"Ix Chel is Lady Rainbow, and she is queen of all the Goddesses. It is she who watches over healers and helps them. She also makes medicine plants grow and leads us to them. She is the guardian of all the forest plants and queen of the forest spirits who guard the plants and animals. She is also a friend to the healer."

Don Elijio continued working toward the root with his pick. "What is this root called?" I asked, turning up my nose and making a face that made him giggle.

"This is *Zorillo!*" he announced excitedly. "And it is the largest one I have ever found in forty years of collecting on this mountain. This *cabrón* must be as old as me—a great-grandfather of the forest."

Zorillo means "skunk" in Spanish, and this root certainly deserved its epitaph, Skunk Root.

While his steel tool pecked at the earth, he exulted, "In the name of the Father, the Son, and the Holy Spirit, I take the life of this plant to cure the sick, and I give thanks to its Spirit."

He let out a squeal of surprise when the thickness of the root was fully exposed. "We use this root for many sicknesses. I use more of this than any other plant in the forest. I'll work here to dislodge this grandfather from his bed while you cut away the vines overhead and strip off the leaves."

It would take many days, he said, to finish collecting the "grandfather *Zorillo.*" We would have to strip the bark off the vine and mix it with pieces of the chopped root to be used as medicine. The leaves are excellent for soothing baths and are one of the many Xiv, he said.

I set the larger, woody vines aside and scraped off the outer bark onto a sack he spread on the ground for me. He ordered me not to let even a sliver get away. Every blow of his pick removed a large chunk of soft, black soil, and he easily lifted up several feet of ropelike roots. The coarse odor of skunk hung in the air while he piled up the exhumed roots, which I scooped up and emptied into a bag.

We worked steadily for three hours, disturbing the ants, spiders, and snails who lived in the dark earth beneath the roots. I held back the flood of questions building up inside me. I didn't want to disturb the rhythm of his work and the quiet of the forest.

"Learn this plant well, my daughter, as we will fill many *trojas,* corn bins, with this medicine. Its special blessing is that it cures many diseases. Mostly we will use it for *maldad.*"

"*Maldad?*" I asked, while helping him remove a stubborn piece of root lodged under a boulder.

"Later, later, you will see," he quipped.

We gathered up our sacks now stuffed to overflowing with roots, bark, and leaves, and hoisted them onto our backs. We had to adjust our loads several times to make room for tools and machetes before starting down the hillside. Still, he pointed to another vine he said was imperative to collect this morning. He stopped, put down his hefty sacks, and began cutting.

"Here is *Chicoloro.* Very important medicine. Remember it well."

Staring at the vine twisting its way to the forest canopy, I could see nothing remarkable about its gray bark or green leaves. How was I to remember this particular one when it looked just like a hundred other vines around us?

As if reading my mind, he showed me how the vine and the branches form a recognizable cross. This was a sign, he said, to the healer that this vine is powerful but dangerous medicine.

I helped him pull down a twelve-foot piece, which he chopped into two parts, draping one half of the snakelike vine around each of our necks. Soon we looked like human burros making our way down the hill with these loads. The journey was getting difficult with the weight pressing on our backs and the emerging midday heat. The inviting coolness of morning had given way to a mean, blistering sun.

He showed me more grayish, woody lianas, giving each a name, calling it male or female, and providing few explanations about its uses.

As we lumbered past the watermelon patch I was sure we were done collecting. To my astonishment, he handed me another empty sack.

"We must collect Xiv for today. We'll find the rest of the nine plants along the trail back to the village, where patients are no doubt already lined up."

Collect seven more plants? Luckily he didn't notice the stunned look gracing my wet, flushed face, as he continued grabbing and cutting as his load tugged against his wrinkled forehead.

"This is *Cruxi,* Cross Vine," he said, holding up another snippet of green. "You see how this leaf makes a cross over the branch of the vine? Watch for that. When a leaf crosses a branch it is a sign that the plant is blessed with medicinal powers."

I leaned in to get a closer look, but with one bulging sack tied to my back, another tucked under my left arm, a pick dangling off my right shoulder, and a circle of *Chicoloro* wrapped around my neck, it took all my energy just to see clearly. Yet he was quite serious about collecting more leaves, and I was in awe at the amount of hard labor he was capable of. I began to fully comprehend what I was in for as the *zampope's* apprentice.

How I wished for a pen and paper, a tape recorder, a camera, another brain. The extent of his knowledge about the forest plants was staggering. I began to feel overwhelmed at the imposing task before me.

As I continued trailing behind him and collecting Xiv, I also prayed and tried to count how many leaves we had collected. Four more to go before we reached home. I was drenched in sweat, with sharp roots digging into my back, and I was also thoroughly confused, unable to remember even one of the plants he had showed me. Don Elijio, on the other hand, was full of pep, as if he'd just had a nap.

"Over there, Rosita, behind that tree and up that hillside. Do you see the large leaf? Go cut nine for today."

Obediently, I put down my sacks, removing them one at a time along with the liana around my neck, while wondering how a man who saw people's faces as hazy masks could spot a particular plant clear across the road *and* up a hill. I pulled out my machete and headed for the steep hill soaring upward from the road. I looked up to see bunches of shiny, thick-ribbed leaves almost four feet long, growing on a rocky incline and being held in place by a large ball of stringy roots. "That one?" I asked, yelling over my shoulder.

"Yes, yes, that's *Xiv Yak Tun Ich,* Pheasant Tail. Very good medicine for rheumatism," he shouted.

I returned with my sack spilling over, prompting him to tie it up with a strip of green, flexible vine he dug out of his pocket. Without a word, he handed me another sack, empty and beckoning for more leaves. I bit my lip and fought back the tears, determined to keep my promise that I would work hard if he agreed to teach me. Since he was keeping his part of the bargain, I wasn't going to wimp out now.

"Here is a blessed tree," he buzzed on, oblivious to my struggle to forge ahead. "*Eremuil,* Wild Coffee, is its name, and it must always be included in every mixture of nine leaves for the herbal baths. We call it *Che Che Xiv* or chief herb because it is so wonderful. Remember it well. You collect that bush and I will work on this one next to it."

I reached out to pull leaves off the branches, maneuvering to keep my loads balanced, when he reminded me, "You have not remembered to say the prayer of thanks to the Spirit of the plant. The Spirit of the plant will follow you home to strengthen your healing, but only if you remember to give thanks. Otherwise, it will stay in the earth and your medicine will not have power. Listen and learn, child."

It was well past noon, and my water bottle and stomach were empty, but I kept my complaints to myself.

"Ah, this darling little plant is Cancer Herb. We will put that into the formula for today. As a powder it is good for diabetic sores and boils.

This little one with the fat flower is the female and growing right next to it is its lover, the male. You see how tall and thin the male flower is?"

I didn't see the difference at all. When I bent down to pick the leaves, my back load shifted and fell right over my head to the ground. When I reached out to pick it up, the vines around my neck caught in my hands, and soon I was a hopeless tangle of vegetation. Don Elijio gave me a sympathetic look but made no attempt to assist me, only looking overhead at the mounting sun and admonishing me for holding him back so much this morning.

Where were we now? How far was it yet to the village? Were we on the sixth, the seventh, or the eighth leaf? Was this one *Anal?* Is this *Cordoncillo?* I felt like dumping all the sacks on the ground and taking a snooze under the nearest shade tree. I began snickering at how funny I must look, loaded down like an overused coatrack in the middle of winter.

"We shouldn't have done so much shopping this morning at the marketplace, *maestro!*" I said capriciously, trying to get my perspective back.

He laughed and shook his head, saying, "Yes, but we are not done. If we don't collect enough Xiv this morning, I will have to return after lunch, and I'm too busy for that today."

He kept showing me more leaves, more branches, and more vines until my mind was crammed with a prickly wad of knowledge. When I tried to recall even one of them, I drew a frustrating and embarrassing blank.

The end of the road veers uphill for the last mile to the village. At last, he stopped and swung around to look at me. He eyed my sacks and the sweat dripping off my chin and said with gentle crinkles forming around his eyes, "You did well today, my daughter. Do not despair. Day by day, step by step, little by little, I will teach, and you will learn."

CHAPTER EIGHT

Red Gumbolimbo, Naked Indian   Palo de Turista
Chaca   Bursera simaruba

The shaggy red bark of this common tree is one of the most
versatile of all the traditional medicines. The bark is peeled and
boiled as a bath for all manner of skin conditions, including
contact with toxic plants, sunburn, measles, and rashes. A tea
made of the boiled bark is used for internal infections, urinary
conditions, and to purify and build the blood. The resin of
the bark contains an antibiotic substance.

———————

Spending three days a week away from home meant leaving Greg alone
with the farm's seemingly endless, daily chores.

When I'd come home from San Antonio, I could tell immediately
how he had fared. If his cheeks were pink and he stood tall as he waited
by the riverside to fetch me in the canoe, he'd say, "For the chance to
study with Panti—this is worth whatever it takes. Go for it, babe." But
when he was slouched and his face grayishly pale, he'd snap, "While
you've been out on a joy ride with Panti, Rose, I've been here stuck in
the mud and alone."

Greg found it hard to wake up each day and face a new crisis requir-
ing a carpenter, a plumber, or a gardener—professions he was learning
through painstaking trial and error. In addition, there were days when
he alone scrubbed our laundry on a washboard, hung it out to dry, fixed
meals, tended Crystal's needs, and loaded the wood stove with firewood
and kindling.

"I wanted to be a doctor, not a fix-it man," he'd gripe when he was
tired.

After all this time we were still trying to push back the jungle from
around our two huts. It was a never-ending project to clear the under-
brush and chop trees. The sunlight had to penetrate or the pervasive

dampness of the jungle would reduce everything we owned to crumbly balls of green mold.

The burden of a jungle homestead was almost too much of a strain at times. In those early days of our marriage, we didn't always get along and we had our share of arguments. The added stress of my absences made it worse. There were times when he lost heart and I was absolutely exhausted from working full steam at two households.

I felt as if I lived a double life and the transition from one environment to the other was overwhelming. No sooner would I lay my head down on a freshly laundered pillowcase and feel the cool night air move through our open thatch than my mind would leap back to my hammock in Don Elijio's cramped and stifling cement house.

Greg would patiently listen to my worries and triumphs at Don Elijio's: the backbreaking work, the patients' stories, the marvelous tales of healing plants. Rehashing my San Antonio days out of context gave me a sharply different perspective. I'd remember, again, how important it was for me to continue my work with Don Elijio. Speaking about Don Elijio also helped me come to terms with the extent of my involvement with him and my ever-growing love and respect. The venting allowed me to readjust to life on our homestead and the embrace of Greg's arms.

I always came back to the farm caked with mud and covered with bug bites. Greg often joked about burning my clothes and shoes while I was showering for the first time in days. Then I'd climb into bed, naively thinking that a hot, soapy wash in the new shower that Greg had installed had killed off the jungle critters. But the ticks were embedded in my skin and the chiggers jumped from my body to Greg's. My condition couldn't be helped. It was part of my life in San Antonio to dig in the earth and pull out roots and collect leaves that were infested with biting ants and insects.

Once I had a run-in with a toxic caterpillar. I was on a mountain looking for plants with Don Elijio when I closed my palm around its hundreds of tiny, poisonous hairs coated with noxious toxins. Don Elijio rushed to my side when he heard me moan. He grumbled, "Very bad, very bad," while pointing to the furry, elongated creature responsible.

"Does it kill people?" I asked, between gulping sobs.

"Sometimes. But there is no cure, only prayer and time."

My palm felt like it had been plunged into the red hot core of fire. I couldn't pull it away from the flame; the excruciating pain felt as if it was consuming my hand. Don Elijio sat me down under a shady tree. He gently held my wrists, blew on my fiery palm, and whispered Mayan prayers.

I held my throbbing arm to my chest and let the tears flow freely as Don Elijio and I made our way down the mountainside. By himself he carried the sacks filled with the day's harvest. The pain increased with every step.

As we walked, he tried to cheer me up. "If you're going to die today, you might as well marry me," he suggested.

His comment infuriated me. I was scared and didn't want to be teased. Through my tears, I told him he was completely shameless. When we got back to the hut I said I wanted to go home to my paramedic husband.

Don Elijio showed little sentiment, which made me angrier. But I guess he'd seen too many bush accidents to get emotional over my less-than-fatal injury.

I left for home, and by the time I stumbled and winced my way through the forest trail to the river, I was nearly hysterical. Greg was sipping a beer with Mick at Chaa Creek. He was startled by my surprise homecoming and quickly offered the sympathy I was craving before putting me to bed. He went into our stash of homeopathic remedies and gave me Arnica pills, then rubbed Rescue Remedy into the wound. I lay down on the wicker couch as Crystal prepared cool compresses. Then, as Don Elijio had said, we could do nothing else but watch my crimson and blistering palm slowly heal itself.

My absences were also hard on Crystal, now ten. It was a delight to see her when I would return home. She'd run to me and give me an update on everything that had happened while I was gone, recounting tales of tomato bugs, visiting iguanas, and the escapades of our live-in cats, one of which showed her love by bringing Crystal a fresh decapitated lizard every afternoon.

Crystal was often lonely, and for a while she moved in with the Flemings and their children. As the only two homesteads for miles, we were one big extended family. We shared a common goal of carving out a wholesome, nature-based lifestyle. The children were inseparable, and Mick and Lucy were our closest friends.

Without much hired labor between us, we often had to help each other fight back the encroaching forest, which never stopped threatening to reclaim its ancient domain. We offered each other support when one of us ran out of fresh food, supplies, or a sense of humor.

Together we worked to make our tropical existence safe for ourselves and our children. Greg's paramedic experience was invaluable, as was my increasing knowledge of home remedies. When Crystal, Bryony, Piers, and Gonzalo all came down with tropical measles—much more dangerous than the northern strain—I bathed them in a tub filled with boiled Red Gumbolimbo bark. It soothed their painful rashes, lowered the high fevers, and helped them fall into a deep, restful sleep.

In the beginning of my apprenticeship, I would return from San Antonio and search my backyard *farmacia,* pharmacy, for a specific plant, never certain I could identify it without Don Elijio's help. After about a year and a half of tramping about in the rainforest, plants began to stand out as individuals with uses and exciting histories. The dark, mysterious forest of trees and lianas was becoming a familiar place of knowledge and healing.

Once I felt sure that the plant in my hand was what I believed it to be, it took much more time before I felt completely confident that I could use the plants I collected on my patients. Only gradually did I became proficient at using the cures Don Elijio had taught.

Our lives were still difficult and challenging, but our perseverance was showing results. Our natural healing practice in San Ignacio Town was flourishing. As Greg and I began to realize that we had purchased thirty-five acres of medicine—enough to last several lifetimes—we no longer needed to import herbs from the United States.

An increasing number of our patients were coming from the Mennonite community of Spanish Lookout, about ten miles northeast of San Ignacio. The German Mennonites, a Christian sect, had settled in Belize in 1958 to form a farming community of several hundred settlers. They came to us mostly for naprapathic treatments. Hands-on healing had always been a part of their culture, and they arrived at our office by the truckload. We treated all ages, from newborns to the elderly. Sometimes as many as twelve members of one family filled our waiting room.

Gradually, our ugly, burned-out clearing in the jungle was transforming itself into a showplace of tropical flowers, including hibiscus in four shades, orchids, and heliconia.

The fruit trees were beginning to bear their mangoes, oranges, lemons, and avocados. The pineapples and bananas were fat and sweet. We discovered wild fruit trees in our jungle, including *annona, hog* plum, and *zapote,* a favorite fruit of the ancient Maya. Its delicious flavor reminded me of a peach embellished with a dash of cinnamon.

Our organic gardens were sprouting some hearty, deep green varieties of lettuce, with succulent, tender leaves. Collard greens had become one of our most reliable crops, and we relished them in their raw and cooked states. Ground crops such as cassava and *macal* (taro root) had become our staples.

With our finances improving, there were times we were able to hire farmhands and a housekeeper to relieve some of the burden of the relentless, daily chores. We also paid Panti tuition—a source of income that never failed to surprise him but one that he greatly appreciated.

We still had bad days, but more often Greg, Crystal, and I appreciated that we had our very own piece of paradise. There was no more talk of moving back to Chicago. We no longer wanted to be anywhere but at our little riverside farm, which we decided to name in honor of the Maya Goddess of medicine.

We celebrated the christening of our homestead with a bottle of champagne we had chilled in the river.

We named it Ix Chel Farm.

CHAPTER NINE

Anal   *Psychotria acuminata*

A common herb used primarily in the Nine Xiv formula for
herbal baths, for treating skin disorders, swellings, bruises,
nervousness, and insomnia, and for all children's diseases.

———————

One day as we collected *Anal Xiv*, I noticed that Don Elijio seemed sad
and distressed. He hadn't made a single joke all day long. We were on
our way back along the logging road when he cleared his throat and
began to talk. He told me about a woman he called La Cobanera.

She was a Latina woman named Claudia from Cobán, Guatemala,
who now lived in a village ten miles south of San Antonio. She had
come to him as a patient with a pain in her belly. After she was cured,
they began a courtship.

But things weren't going well. La Cobanera had promised many
times to come and live with him: "Until you bury me or I bury you,"
she would say. But the last time he had seen her, she had broken this
promise with yet another transparent excuse.

"I told her she is like the flower called *Amor de un Rato*. Love for a
while. She says she still has much work to do before she can stay with
me. It is always 'next week, next week.' I guess next week never comes."

He had spoken his heart to her, admitting to her he was lonely and
in dire need of a woman to administer his house. "I promised her she
would not suffer with me. I've eaten a few beans in my time and I know

how to make a woman happy. Before she could ask me, money would be in her hand. I have no rushing river of money coming in, but I have a reliable drip."

He was getting visibly more upset with each disclosure, and soon the story rushed forth like an emotional hurricane.

"Each time she comes to stay, she asks for money. The nights she spends with me are very affectionate. But in the morning, it hurts me when she packs up her bag and then holds out her hand. At first it was twenty, then thirty, then she asked me for eighty dollars. Rosita, she's playing with me. Flattering me, loving me in the night, and laughing at me in the day."

He said a patient had told him that La Cobanera was seen in Belmopan on the arms of another man. She was known by many names: *Limpia mundo,* meaning one who cleans out the world; and *Rastrillo,* for one who rakes in, in reference to her voracious appetite for money.

"Watch out for her, old man," the patient had warned. "Don't believe her lies, for she is not what she seems."

Panti had confronted Claudia and informed her he knew of her capricious habits but that he was willing to forsake the past. "She can have a fresh start with me. I'll take good care of her just as I did my Chinda."

It pained me to see him court an unappreciative woman. Such a good and attractive man shouldn't have to bargain with a woman for her affection. It was apparent that he had a trusting nature, a trait we both shared. Perhaps it's a function of our healing gift to be too accepting and forgiving of human frailty.

Finally, I blurted out, "She shouldn't do that to you, Don Elijio. She doesn't realize what a wonderful catch you are."

"*Mamasita,* it stings me when I must plead with her like a schoolboy not to go," he droned on in a sorrowful voice.

We trampled along the road with our wet sacks sinking us into mud puddles and potholes that got deeper and slushier with the steady drizzle. I kept looking over at him to see if his mood was getting any lighter, having vented some of his despair. I was getting more protective of him with every passing day, and I was angry. I was willing to devote so much of myself to this lovely man, whom I deeply respected, while another woman mocked his search for a true companion by teasing his lonely, aging heart.

A few weeks later, I met La Cobanera for the first time. It was just after the last patient had left and I was outside cleaning up while Panti was resting after a naprapathic treatment. I heard an engine shut off and a car door slam just outside the gate. I anticipated the face of a patient turning the corner in search of Don Elijio.

But a short, stocky, barefoot woman dressed in patched and torn clothing appeared and plopped herself on the stoop. She twisted around and peered around the room until her gaze landed on me. "Who are you? Are you sick? And where's the old man?" she said in one breath.

I told her who I was. "Who are you?" I asked.

"Claudia, at your service. Maybe the old man has told you about me, no? I've heard a lot about you. I can tell he likes you. So, what does he say about me?"

While she drew a breath, I mulled over her indelicate question. My response was snagged somewhere between unrestrained frankness and polished tact.

"He says he really likes you but wishes you would stay with him and not leave him each morning to go home."

"But I have to. I have a house, five sons, a cornfield, and animals to care for. I'll stay with him when I'm ready, but he has to help me. This takes money, and only with money will the dog dance. *Con dinero baile el perro.*"

We heard stirring in the bedroom, and she hoisted herself up and peered behind the curtain to his bed.

"So, you're back!" I heard him say, with a detectable chilliness.

"I can stay a week this time, my king. My corn is all harvested, and my sons are storing it away. The youngest will make meals for the others, and they'll manage fine without me," she said matter-of-factly.

Panti pushed the curtain aside, still pulling on his cotton shirt and scratching his bites and stings from the bush.

"This is my student, Rosita," he said, motioning toward me. "She'll be here too for a few days helping me collect medicine. You can get to know each other." She turned toward me, and her smile reminded me of a Cheshire cat's.

She stashed her bags in his small room, then sauntered over to the kitchen hut, where I heard her stoking up the fire. Panti and I soon returned to the hut to resume our chopping chores.

Having another woman there—and one who knew him so intimately—gave the hut a cozy feel. While she stuffed papers in the fire and fanned the nibbling flames, she gabbed effortlessly to a sour-looking Panti. We heard about her corn, her unemployed sons, and a horse that had foaled in her yard. But within the half hour, Panti began warming up to her, failing miserably at his efforts to remain aloof. She seemed the sort of person who was oblivious to such subtle rebukes anyway.

He began weaving tales that were a keen match for her ambitious ramblings. I watched their faces light up from chuckling and simple, wide-eyed wonder at any morsel of news about livestock, crops, weather, history, and child rearing. I was reminded of those wind-up dolls that keep chattering away until they fall over.

I tried giving them some privacy by concentrating on my work and remaining relatively silent, but I was bowled over by his animation in the presence of such a romantic guest. If she was sincere, it would be a prayer answered, I thought—a companion to share his overburdened days and long, lonely nights. If she were using him for his money or reputation, then I was witnessing a budding tragedy.

They talked all afternoon, through dinner, and into the evening. I became as absorbed with her stories as he was. She spoke a different Spanish than his, and her dissimilar background and life history fascinated me. Soon she was disarming me with her charm as I too joined in the talkfest. When I asked about the father of her five sons, she said without flinching that he had left her many years ago. She had raised the children by herself. It was a pitiful tale of hardship not unlike those of many Central American women I'd met.

As we prepared for bed, I offered to hang my hammock in the kitchen hut, but Panti insisted I stay in the living quarters, where I slept just a curtain away from him and his new *amante*. I heard them cuddling together in his large, denim hammock, whispering and giggling until three o'clock in the morning, about the time I faded into a deep, numbing sleep.

At first light, he jerked my hammock as usual: "Child, wake up. The time is right." In spite of a busy night, he seemed bright-eyed and full of energy as we took off to his *milpa*. I could barely keep up with him. Although I still had serious doubts about La Cobanera, I was convinced, if nothing else, that she was an enthusiastic companion for him. I hoped that he had the power within him to convince her to stay.

70

CHAPTER TEN

Wild Yam   Cocolmeca   Dioscorea sp.

The starchy tuber of this thorny vine provided the base molecule
for the birth control pill. Traditionally, the tuber is boiled
and drunk as a tea for rheumatism, arthritis, anemia,
kidney complaint, diabetes, and as a blood purifier. It is
rich in iron and minerals.

One day, out of the blue, Panti said, "It's time to do a *Primicia,* Rosita. I
want to introduce you to the Maya Spirits."

I was surprised and honored. The *Primicia* was an old Maya ritual
that had all but faded from modern daily life. The purpose of the ceremony is to give thanks to, worship, and ask favors of the Nine Maya
Spirits . . . and God.

Panti talked as we tried to loosen a thick, wrinkled root of Wild
Yam.

"When we were young, the villagers did a *Primicia* in the Catholic
church after each planting," he said. "We stayed up all night, chanting
and calling out to the Spirits for rain." Before dawn, torrents of rain
would pour, announced by loud cracking thunder. The thunder would
awaken the tender seedlings in the fields.

"We did nine *Primicias* a year for planting for harvests, for rain, for
sick people, and sometimes just to show our love for the Spirits. For
crops, the farmer would take the blessed *atole* and pour it on the four
corners of his *milpa.* Hunters would do a *Primicia* after killing nine deer,
saving each of the jawbones to place on the altar. Then the deer would
lie down for you to shoot them for food."

Without the Spirits, he said, life would be impossible. It is they that bring the rain, the thunder, and the seasons and cause all things to thrive and grow. They are the caretakers of the world who look after the people, the animals, the plants, the harvests, the seasons, the day and the night, the crossroads, women in childbirth, and all aspects of daily life. Each Spirit is entrusted with certain aspects of life, he explained. For example, Yax Tum Bak is the Lord of the plantings, and Chac is the Maya God of rain.

There were nine Spirits. For this reason, nine was a holy number for the Maya, he explained while we collected leaves of the thorny *Escoba* palm. "I can't tell you all the names of the Spirits because you have no sastun and it would weaken their powers.

"People rarely honor the Spirits anymore," continued Panti sadly as he stuffed the leaves in his sack. "They have no respect for the Lord of the cornfield. If you ask them to honor the Lord they will only laugh at you and say, 'What Lord of the cornfield? I'm the only lord here. I don't believe in your old Spirits, old man. We don't need them anymore.' But now look at the ugly corn they harvest and the droughts. They need the Spirits more than they know.

"And our Spirits need the *Primicias*. It is through the *Primicia* that they are invited into the world of mortals. It is through the *Primicias* that our prayers are answered and the Spirits have life."

I had long been curious about his Maya Spirits, which he also referred to as *Segundo Dios,* second to God or the Right Hand of God, and I asked what they looked like.

"We don't see them as people with faces and bodies, but only see and feel their presence in the Winds. They are in and of the Winds and come to the earth in these same Winds," he sang out. "The lightning is their machete. Their backs are visible in the flashes of lightning that tear across the skies during storms. The thunder is the sound of their voices."

With his machete in his right hand, standing tall, he mimicked the sound and fury of thunder and lightning.

"Boooooom, kaboooom, boooooom, kaboooom," he shouted with great stage presence, whirling his machete over his head.

"They sound frightening," I said.

"No, no, noooh, they are very good friends," he exclaimed excitedly. In the Maya religion, God and Spirits intermingle with mortals in every phase of life, he said. The Spirits are almost always friendly except when they see radio, television, and incest. "They get lonely and long to help us with everything," he said. "Whatever we need they are there. If we only ask."

I thought of them as the Oversoul of the Latin peoples, similar to our concept of guardian angels or, perhaps more accurately, the archangels like Michael, Raphael, and Gabriel—the big honchos of the spiritual realm.

Don Elijio said that other lesser Spirits look after every plant and animal. He called these *duenos* or Lords, and like the Celtic elves and fairies, they were elusive and mischievous.

He also believed in the Nine Malevolent Spirits who reside in the nine levels of the underworld.

"They are the ones who answer the calls of the black magicians," he warned. "They also come on the Winds, but they do evil. They like to come in the night."

At one time the Nine Benevolent Spirits had lived in Tikal, he said. But once the archaeologists had come, they had fled seeking refuge in more remote ancient temples as yet inviolate. At first they went to live at Uaxactun. "When archaeologists came to Uaxactun, the Spirits fled to a place called Caxcun on the border of Belize and Guatemala where three hills come together to form one peak.

"Caxcun is enchanted and is now the home of the Maya Spirits, and no one can or should go there," he told me. "Many have tried. All have failed. Some *gringos* tried to climb to the top of the peak and were pushed back repeatedly by the Winds, and the dirt under their hands turned to sand and they could not get a grip. Later, the same *gringos* tried to come back with an airplane to fly over the area, and even the airplane was prevented from going there by a strong Wind that continually blew it back from the area and prevented it from flying directly over Caxcun."

When the good Spirits had left Tikal, he said, the evil ones had taken over Tikal as a favorite earthly haunt.

"That is why I would never go to Tikal again," Don Elijio said. "I am afraid."

Spirits also lived in lesser ruins like Xunantunich, a small ancient city just across the river from Succotz Village where he had grown up.

"When they first opened the tomb at Xunantunich, the workers fell to the ground in a stupor. Just before passing out they heard hmmmm-mmmmmm," said Don Elijio, making a loud and eerie humming sound. "Some of the workers died."

By now we were standing in front of a Guaco Vine, which I found hard to discern from *Chicoloro* until Don Elijio sliced through it and showed me a characteristic starburst pattern at the core of the vine. He held the severed vine under my nose and told me that this was the female of Contribo and used for *ciro*. The odor of Guaco was faint in contrast to Contribo's overwhelming pungency.

As we chopped Guaco, Don Elijio explained that he wanted to hold a *Primicia* for me now because Good Friday of Holy Week was coming and that was the perfect day to meet the Nine Benevolent Spirits.

"It is the holiest day of the year," he explained. "It's the day the Maya Spirits go out visiting their people all over these Maya lands."

I was a little confused. "Why is Good Friday a holy day to the Maya?" I asked. "It's a Christian holiday. What does it have to do with the Maya people?"

"You mustn't believe everything you hear, child!" he answered. "Jesus, Mary, and Saints like Michael, Joseph, Gabriel, Margaret, and Magdalene came to this land of my people many centuries ago. When they came, the nine Maya Spirits called a heavenly council with them. They're not like you and me, you know. They're not jealous or full of envy. No, they all got together at this big meeting and decided to work together for the salvation of the peoples' souls. Together they answer our prayers, heal the sick, and hold our hands when we die."

Don Elijio also prayed to the Four Virgins. "Didn't they tell you of the Four Virgins in Catholic school?" he asked me, exasperated. Patiently, as if speaking to a child, he explained about the Virgin of Carmen, the Virgin of Guadalupe, the Virgin of Fatima, and the Virgin of Lourdes. "We pray to them and they answer with miracles, Rosita. Faith is what moves them to work for us."

As he told me about the Virgins, I realized that the four aspects of Ix Chel, queen of the Maya Goddesses and the mother of all people, had simply been transferred to the Virgins. Ix Chel was the overseer of four domains: as a young maiden spirit, she was in charge of childbirth and weaving; as an elderly crone, she looked after medicine and the moon. The indigenous people of Central America had conveniently cloaked their Goddess by making her four in one and one in four. Four was also a holy number to the Maya. It was like the mystery of the trinity: three in one and one in three.

Don Elijio's religion was clearly a mix of the old Maya and Spanish Catholicism. I asked him if he had heard about the Christian Saints when he was a boy, and he admitted he had not. It was not until the Catholic priests told the villagers they should honor them too that he had added them to his cosmology. I thought to myself, it had been wise of the Maya elders to allow their religion to absorb certain aspects of Catholicism rather than have their own completely obliterated as pagan competition.

I had to admire Don Elijio's simple and potent faith in unseen powers. I also admired the way he involved his Spirits in his daily actions. This was very different from my experience with Catholicism as a child. Religion and everyday life were much more separated in my upbringing.

It was getting close to the noon hour. The day was heating up, and our sacks were full of Wild Yam and *Zorillo,* Skunk Root.

"I brought us each a mango, Don Elijio. Let's sit down and rest for just a bit and enjoy them," I said, showing off the plump, golden red fruit we had harvested that year off our young trees. He spread out a plastic flour sack for us to sit on near a flowering *Yax Nik* or Fiddle-wood tree. Tiny, purple blossoms sprinkled the air around us with a faint, sweet perfume.

As we savored the juicy bites, the subject switched from Spirits to La Cobanera. Recently Claudia's jealous and self-consumed nature had become all too evident. There had been outbursts with female patients; I had watched her on several occasions interrupt his abdominal massages of women. Jutting her angry face behind the curtain, she'd demand, "How long will this take? Do you have to massage every woman who comes in here?"

Panti was aghast at her performance, and the patients didn't like it much either. Her suspicions could inflame whatever doubts his female patients harbored toward him based on the street gossip that had initially made me afraid of him. He was also gravely insulted by her accusations, as he especially prided himself on his flawless code of ethics with women patients. Often he told me, "Rosita, we cannot sin as easily as our brothers. It is a grave, mortal sin for us to contemplate harm or to disrespect our patients. We lose the help of God and the Nine Benevolent Spirits if we harm our patients. No, it is a sin, and I don't sin. If I sin against humanity, only the Nine Malignant Spirits will want to work with me, and I'm not interested in harming people, only healing them."

Claudia was jealous of me as well, hinting that Panti and I were lovers, hiding our affair from her and Greg. We only laughed at that, realizing the pitiful extent of her paranoia.

In the bush that day, I advised him, "Papa, you're a grown man with lots of experience in life, but this time I fear your loneliness and need for a woman has blinded your judgment. How could you possibly do your work in peace with that scowling face peeping behind the curtains? Just think of how she sees our relationship. She sees mud where waters run clear. No good, *papasito*. No good!"

"She could change. Maybe she only needs a good man to love her in spite of her bad ways. I see all that you say, but I can't help myself." I couldn't help but wonder if Panti was seeing too much good in her. I had the same habit of being too trusting. I remembered what my Assyrian grandfather, Simco, used to tell me, "Honey, you listen to papa. You so good you crazy!"

Panti leaned over on his pick for a moment, wiped his brow with his stained, embroidered handkerchief, and sighed deeply. I half expected to see tears fall from his eyes. Instead, he only shrugged and said meekly, "Well, I've already invited her to come to your *ranchito* for your *Primicia*. Maybe with some time together away from the village gossip and my grandchildren, we'll have a little lovers' escape. When she sees my work and my gift, I think her heart will be softened. She'll become the good woman I know is hiding behind the shrew."

Since she had never done anything unselfish for him, he planned to test her devotion by asking her to scrub his laundry in the river while they stayed at our farm.

I was thrilled at the notion of having him become part of my household for three days, which was as long as he dared disappear from his patients.

On the way back to Ix Chel Farm that afternoon, I thought about how Don Elijio, a man with the power to work with spiritual forces, was also completely vulnerable. He was a frightened old man feeling unloved, misunderstood, and abandoned. In love he had the same problems as everyone else. Somehow this made me trust him all the more.

As soon as I got home, I announced my news of the forthcoming *Primicia*. Everyone was as excited as I was. My son James, now twenty, was staying with us during a break from college. He and Crystal were thrilled by the idea of an ancient ceremony in their mother's honor.

The thought of having a *Primicia* as a thanksgiving was quite fitting, as we were celebrating an anniversary of Ix Chel Farm that week. We knew we had a lot to be thankful for: we were still alive and together, none of us had been bitten by a snake, we had a bit of fresh, organic food on our table, and Don Elijio was sharing his knowledge.

Greg and I decided to build the altar for the *Primicia* on the ancient Maya mound we had uncovered during construction. There we had found stone tools, spinning whorls, shards, and obsidian blades, evidence of an earlier river community—just twenty feet away from our kitchen hut.

Life was undeniably easier now. We expected more hardships, but we feared them less, having mastered some knowledge of jungle lore and survival tactics. We were eager to thank the Maya Spirits for whatever role they had played in our good fortune. We could hardly wait for Holy Week, *Semana Santa*.

Tzibche   Crotolaria cajanifolia

Somewhat of a rare herb, *Tzibche* is used in the treatment of
many spiritual diseases and as a protective brushing before the
sacred Maya *Primicias.* It may also be collected as one
of the Nine Xiv formula for herbal bathing.

Good Friday broke with seasonal dreary rain and fog, which had been
soaking our farm most of that Holy Week. Although the farm was now
blanketed with young blades of grass, there were still many patches of
slick mud. We had been living in knee-high rain boots for days and
bathing in rainwater under the outdoor shower. We'd also kept the
wood-burning stove lit all day to keep ourselves warm and to dry the
laundry hanging everywhere in our kitchen.

We had homemade muffins, mangoes, and Lemon Grass tea for
breakfast that morning around a table Greg had fashioned from secondhand mahogany boards. But this was no customary, workaday Friday:
This was the day my family and I were to meet the Maya Spirits.

We began scurrying around in preparation not only for the *Primicia*
but for our special guests: Panti and Claudia. James left to canoe across
the river and walk the mile to the road to guide our guests to the farm.

Soon I heard James yelling from the riverbank, "Mo-o-o-om. We're
crossing over." Within minutes we saw Panti slowly and deliberately
climb the hillside steps, which Greg had just finished building a few
weeks before.

Behind Panti was a red-faced, puffing Claudia, bearing his sack of dirty laundry on her head. James held up the rear, oar in hand with a smile as wide as the Macal River on his handsome face.

We exchanged warm greetings, and Panti marveled at my large kitchen, with its thatch roof soaring twenty feet above our wood stove. He was used to small rooms with few windows and imposing darkness.

I led them to the guest room—a rough and rustic, unpainted frame house with a thatch roof, but they let out squeals of approval as if it were a four-star hotel. Panti put down his plastic flour sack, stretched his slight frame out on the cushiony bed, and inched his body around like the hand of a clock.

"It's so big and soft. This is the best bed I have ever seen in my life. Rosita, you treat us like king and queen." Claudia grinned, nodding in agreement, quickly spotting the rocking chair with a pillow seat. She squatted her plump frame onto it and began rocking and sighing from fatigue.

Panti declined breakfast, saying it was his custom to fast on Good Friday until dusk. I noticed Claudia was more tender with Panti than I had ever observed before, and I began to think perhaps he was right—she just needed a good man to love her and draw out the dormant angel within.

Once in our kitchen amid the bloomers, overalls, and towels hanging from the rafters, Panti appropriately took up court in our most stately chair. He was kind and patient with our endless questions about the *Primicia* and its significance, enjoying his role as both priest and teacher. As we talked of Maya Spirits, we were startled by a loud, frantic call from across the river.

We shouted back and heard a man yell, "Is Don Elijio there with you?"

"Yes," I hollered reluctantly, fearing that Panti would be whisked away by a supplicant who had tracked him down. I knew if an infirm or needy person came looking for him, he would tend to that patient and forsake his respite and our much-anticipated ceremony.

"We need to see him. It's urgent! Please let us cross!" came the voice again.

Panti only laughed and rocked faster in his regal chair. "You see what I told you, child. They follow us everywhere."

"You haven't even been here an hour and already they're calling for you," I said, handing an oar to James, who went down to the river to ferry the visitors across. A few minutes later James reappeared, with two men following. They looked desperate.

"We went to San Antonio just after you left, but your granddaughter told us you were here," one pleaded. "I need to speak to you in private, *tato*, please. It's urgent!"

Rising from his chair ever so slowly and stiffly, Don Elijio grabbed my outstretched hand and held on while we gingerly walked across the slippery grass to his little honeymoon cottage.

After they departed, I went to the guest hut, where he told me that one of the men had been involved in a serious crime in San Ignacio, where a court hearing would be held the following week. He had asked Panti to enchant the judge and the jury in his favor. I was shocked to know that he could or would do such a thing, but he explained that the charm only works if the defendant tells the absolute truth. If he has lied, the charm is broken. The special prayer for court cases ensures that the defendant is set free after telling the truth because the judge or the lawyers can't find a document crucial to the conviction. Thus the case is dismissed.

Then Panti did something else that surprised me. He brought out the flour sack he'd carried with him from San Antonio and whispered, "This bag has my money in it. Hide it for me until we leave."

I carried the bag into the main house wondering where to hide it. Peeking inside the bag, I saw stacks of neatly piled and rubber-banded bills in every denomination. I shook my head in amazement and stashed the little bag on a shelf over our bed.

Distractions aside, I went back to the kitchen hut and helped Claudia stoke up the wood stove to boil the blue corn. The setting sun signaled that the start of the *Primicia* was fast approaching.

Panti tested and retested the boiled corn until he was sure it was the right tenderness. Claudia had never participated in a *Primicia* before this day. Her Guatemalan village had completely converted to Protestant fundamentalism when she was just a baby in the 1930s.

She and I put the boiled blue corn through a hand mill and reboiled it with water and brown sugar. On the altar, we arranged nine white flowers—one for each of the Nine Maya Spirits—and vases of flowers and bowls of fruit. Two wooden angels I had brought from Mexico graced either end of the altar with their delicate wings outstretched and their hands folded in prayer. Four candles were burning under umbrellas, and beneath the altar were clumps of the sticky, resinous Copal incense smoking on a bed of coals. Its rich, spicy smoke swirled up and billowed in the damp winds, as a light drizzle fell.

Panti and I went to search for *Tzibche* leaves for protecting the participants. We found them just yards into the jungle on our property. He marveled at how much medicine we had growing wild behind our huts.

Returning to the altar, we placed photos of loved ones and people with health or spiritual problems whom we wanted to be blessed by the Maya Spirits. All of our crystals came out of their little velvet bags and were placed on the altar. We worked quietly with a sense of reverence, as an air of hushed spirituality permeated the farm even before the *Primicia* began.

When every detail was complete, Panti summoned us to stand around the altar. I brought the *atole* from the kitchen and Claudia handed him his bag of *jícaras,* gourd bowls. Tenderly, he placed the nine bowls on the altar. He used the tenth bowl, called a *julub,* only as the serving tool, filling the others a third of the way with the corn *atole.* He was portioning it out as a mother would the last remaining nourishment to her hungry brood.

He turned to see that all were present. Greg, Crystal, James, Lucy, Claudia, and I stood around him, holding our breath for the Maya mass to begin. I could hardly contain my excitement. I felt privileged to be standing next to my *maestro* at this candlelit altar at the edge of the Maya rainforest.

With the *Tzibche* branches in his hand, he brushed each of us, forming nine crosses on our bodies. He explained that this was to protect us from the powerful presence of the spiritual Winds. They do not intend to do harm, he assured, but sometimes if people are weak, sick, or in a negative emotional state the presence of the Spirits can make them quite ill. But the brushing with *Tzibche* was sufficient protection. He murmured a Mayan prayer, with only "God the Father, God the Son, and God the Holy Ghost" heard in Spanish.

After the brushing, he motioned for me to stand at his side in front of the altar. I took my place, feeling like a child taking the stage, nervous about remembering my lines. He made the sign of the cross over his chest, held his hands out in front of his body, palms facing outward, and began the lilting prayer of the sacred *Primicia*. The rest of us were completely drawn into his presence and power. His words were unintelligible to all, but the emotion, love, and faith of this man needed no translation.

I was electrified standing by his side, feeling each word take flight into the air around the hillside mound. Greg placed more Copal on the coals, and we savored the pungent sizzle summoning the Spirits to come inhale and listen to Panti's holy chants. Tilting his head back, he sang out what sounded like a shrill bird call, "*Ki-ki-ri-ki-kiiiiii.*" He continued chanting until he bowed his head and made another sign of the cross. While waiting for the Spirits, it began to rain.

Sheets of water washed across us, forcing Panti and Claudia to seek shelter in the kitchen, still watching the altar over the half-wall. The rest of us braved the rain under slickers and umbrellas, not wanting to break the spell of this singular, charged moment.

Within the hour, the storm gave way to an eerie stillness, and Panti resumed his melodic *cántico*.

I was honored to hear that he mentioned my name with the respectful prefix of *Doña* several times, then he gestured with his outstretched hand in my direction. I still found it hard to believe that I, Rosita, an Assyrian-Italian woman from Chicago, was standing in the middle of the jungle with a Maya priest, being initiated into the realm of the ancient Maya Spirits.

Panti uttered the last phrases and was about to turn away from the altar, when he froze on his feet. He peered slowly over his right shoulder, and we all followed his gaze. The air was heavy and still; the palm fronds hung motionless, as if they were posing for a still life painting. Yet the flame of one candle bent three times toward the west, while the flame at the other end of the altar bowed down three times in the opposite direction.

I glanced at Don Elijio. He only smiled and said, "They're here right now. They've come. They have heard our *Primicia* chant and have drunk the spirit of the *atole*. Look there," he added, pointing with his crooked, brown finger. A few yards away from the altar, a lone palm frond rhythmically waved back and forth, as if to signal the Spirits' celestial presence.

We all looked at each other in stunned silence, as if to say, "Did you see what I saw?"

As we quietly collected the offerings and trinkets we had placed on the altar, I shyly asked Panti what he had told the Spirits about me. He stopped and whispered, "I told them that you have a pure heart and that I have accepted you as my beloved disciple. I asked them to give you every consideration they have given me."

"Thank you, my friend," was all I could manage to utter.

We loved the Copal incense so much, we added more resin and carried it around the farm as a blessing. Panti said it would bless each and every corner and shadow.

He instructed Claudia on how to gently wash and dry his beloved *jícaras,* which he had used for hundreds of *Primicias* over the past half-century. When she handed them back to him, he turned toward me with outstretched hands replete with his magnificent gourds.

"I leave these *jícaras* with you today, my daughter," he said, looking at me with fatherly pride.

Tears of joy and gratitude fell from my eyes onto his wrinkled hands. He smiled and grasped my hands in his rough, calloused palms. We had forged a bond of honor to the Gods of healing.

Copal Tree   Pom   Protium copal

The tree and its resin are sacred to the Maya, who burn the
aromatic resin as incense during ceremonies of purification,
thanksgiving, and supplication to the Gods. The burning of
Copal incense is a specific treatment for spiritual diseases such as
envy, fright, the evil eye, grief, and sadness. It is believed to ward
off evil spirits and black magic. The powdered bark is applied to
wounds and infected sores. The boiled bark relieves stomach
cramps and destroys intestinal parasites. The resin forms the
basis of a varnish used on many fine woods.

Shortly after the *Primicia,* a refugee from El Salvador named Orlando
showed up at our gate asking for work. We had a herculean task before
us of felling several dead sixty-foot trees, sawing up the branches, and
burning them so the snakes wouldn't take up residence in the under-
brush. We knew we couldn't do it alone, so after counting our scarce
cash, we agreed to hire him for the next three days.

By the end of the second day, we had felled dozens of trees and were
preparing to burn the piles of branches. I hated clearing the forest,
thinking it sad to see such noble trees lying on the ground, having surrendered their lives to our needs. That day our farm looked like a graveyard to me, with our machetes dripping with the trees' life-giving sap.

We knew it was imperative to set the brush on fire at the appropriate
time of day, due to the wind's ability to carry away the flames, especially
during the dry season. One had to have the experience and knowledge
of the bush to do it right, and after talking with Orlando, he assured us
with supreme confidence that he did. He puffed up his chest and said,
"*Si, señor, yo sé bién.*" Yes, sir, I know well.

At eight the next morning he ceremoniously set fire to a dozen piles
of brush and tree trunks only ten yards from our two huts. Within minutes Greg and I knew he and we had made a serious mistake. The early

morning wind began swirling around the farm carrying menacing tongues of flame. We quickly were engulfed as ten-foot-high flames whipped around in every direction.

Crystal was sleeping on her bed under a window, and a ball of fire blew in just above her head, scorching her pillow and nearly setting her hair aflame. We quickly evacuated her, rushing her over to the Flemings' house. To make matters more dire, Mick was away that day and his water pump broken.

The fire was out of control. We grew increasingly horrified, running to each new flame and futilely trying to exert control over its fiery appetite. The wild coconut palm fronds were transformed into twenty-foot flaming projectiles. We were encircled, and for a moment Greg and I stood paralyzed with fear and indecision. How could we quench the hungry inferno with only fifty gallons of water on reserve in a drum and a useless river pump too weak to lure the water up a hundred-foot hillside?

Orlando showed no interest in helping us put out the fire. As if nothing was going on, he nonchalantly asked for a lift to town. Although Greg spoke no Spanish at the time, Orlando had to have understood Greg's angry shouting. "You crazy son of a bitch! You started this fire and now you want a ride to town while everything we own is going up in flames?" Greg screamed over the sounds of crackling brush, his fists waving wildly in the air. "Get out of my sight before I kill you with my bare hands!"

Orlando strolled off with his two days' pay already stuffed in his pockets. By this time I was totally panicked. Sparks were settling on the delicate, flammable thatches, and if either roof caught fire, the battle was over. I had seen a thatch house burn up in Mexico. Once the roof was in flames, the family lost the house and everything inside within ten minutes, including a sleeping baby.

I began losing hope and started yelling frantically at Greg. He was also panicking but pulled himself together, calling on ten years' experience as a paramedic with the Chicago Fire Department.

He shoved a rake into my hands and told me to start raking out a fire line just beyond the circle of flames. It seemed futile, since the circle was eating up more inches of brush as the seconds ticked by. He started the water pump and quickly began sucking the fifty gallons of water into a hose, which he aimed at the thatch roofs.

"The flames are too big," I screamed. "They're passing right over the fire line."

"Don't you think I learned something about fighting fires after ten years of rescuing people? Just do as I say, Rose, or we're going to lose everything," he ordered with enough force to convince me to trust his instincts.

While my arms and legs worked furiously, taking long, gripping strokes with the rake, my mind and heart concentrated on intense praying, as I pleaded for mercy for our home.

We fought the horrendous, unpredictable fire for nine more hours, never stopping to rest or eat. We saved the houses, but the rest of the farm was left a blackened surface of smoking stumps. Snakes ran out from every direction, slithering only inches from our feet. They had been comfortably nesting under the damp, cool piles of brush until the blaze surprised them.

At dusk, a particularly beautiful sunset in dazzling shades of orange and purple filled the sky. We sat on the one unscathed hillside, wiping each other's faces, drying our tears, and reminding ourselves to be grateful that we were still alive.

We went to join Crystal at the Flemings'. No words could describe what we had gone through that day. Lucy put on some hot water for us to bathe, made us dinner, and listened to our story, which got more palatable with a couple of stiff drinks.

For the next eight weeks, we saw small brush fires ignite in old tree stumps just beyond our clearing. It became part of the daily routine to search out beds of smoldering embers that might burst into flame at any moment with the right wind direction and velocity.

Greg grew more glum with each day, finally giving in to a dire case of what we've learned to call "the Belize blues." It's a syndrome that plagues most newcomers to the untamed jungle. When Greg wasn't depressed, I was.

After the fire I couldn't leave the farm to visit Don Elijio, and I suspended my stays in San Antonio indefinitely. I sent several messages to Panti through his grandson Angel.

Although the fire had obliterated years of work, we returned to some semblance of normal operations within two months, and I felt comfortable about resuming my visits to San Antonio.

Early one Friday morning, I waded across the river on foot, as the water was no more than thigh high at its deepest point during this particular dry season. Soon I was reveling in the forest glen above the riverbank and welcoming the bulging-eyed lizards and prancing scorpions. The jungle can be hostile and unforgiving, but I had a deep, undeniable love affair with it. The aroma of moist, rotting humus was ambrosia to me, and it felt refreshing to be back again on my solitary treks through the beckoning, high bush.

I found Panti at home, sitting in the darkness of his cement house. His shoulders sagged and his expression was lifeless. Something was terribly wrong.

I cleared my throat to let him know I was in the room, then planted a warm kiss on his weathered forehead. He looked up, and slowly a sign of recognition flooded his soft eyes. "Ah, *mamasita*," he said feebly. "Where have you been? I thought you'd forgotten me."

My heart ached to see him so forlorn and hunched over. I asked about the notes I'd sent with his grandson, but it was obvious the secondhand messages hadn't allayed his fears that I'd deserted him.

"I was sure you were never coming back," he said, despite my assurances that would never happen. I relayed the whole, sad story of our fire and what dumb *gringos* we were for placing our lives in the hands of a stranger. He was sympathetic, reminding me that such a bank of knowledge about jungle survival must be learned through fateful mistakes if one didn't learn it growing up here.

"A dry season fire should never be started in the morning. What a lying idiot you hired," he cursed. "A morning fire is caught by the winds and blows all about. It should be done about four in the afternoon, so it can burn for a few hours and the dampness of night puts it out naturally."

Then he looked down at the floor. Something was still bothering him. I asked about La Cobanera and knew I had struck a raw nerve. "She has robbed me," he said, keeping his head down. "She stole my life savings. It happened a month ago yesterday. What an old fool I am, Rosita, trusting that old cow."

She had finally agreed to come stay with him on a permanent basis but first had to go home to pack up her things. They agreed he would send a taxi when she was ready to move.

He waited for two weeks, lying awake at night, thinking and planning their new life together. He went around the village looking for a small house to purchase, thinking she would be happier living in a home not so close to his family and patients. He told Angel that they were getting married, and Angel agreed to welcome her into the family.

She arrived with no possessions, accompanied by one of her five sons, who she insisted would be living with them.

"I agreed. That's how stupid I am. I was so happy to see her, and we kissed and hugged like two lovebirds," he said, almost happily remembering his hopefulness.

The next day she was already packing to leave, saying she had forgotten something important at home. "'Don't go!' I begged her. But she must have had a good laugh at me, while my heart was breaking."

Claudia had been gone for four days without a word before Panti walked into his bedroom and noticed that the lock on the chest housing his money and valuables had been picked. It had been rearranged to give the impression it was intact, when actually it was broken.

"My heart began to pound because I knew what had happened. I opened the chest and start hollering, 'Gone, gone, my money's gone!'" He said Angel came running, and together they stared into the empty chest. Every dollar and every piece of gold and silver had been stolen.

At first Panti thought it must have been a thief, a stranger, until Angel told him he had caught Claudia and her son in the room when Don Elijio had left for the bush to collect medicines. They were acting suspiciously, but Angel hadn't interfered, knowing she was to be his grandfather's wife. Claudia and her son had stayed in the sweltering room for nearly an hour with the doors and windows closed for privacy, Angel said.

Don Elijio and Angel then went to the police, who followed them to Claudia's house. She greeted him with a sneer, saying, "So you would never come to visit me when I invited you, but now you arrive with the police!"

He shouted back, "Traitor! Thief! Your flag is a sack!" meaning "You have no home, no country, and no pride."

He told me he later consulted the sastun about who had stolen the money, asking if Claudia had been the culprit. It answered yes.

When the police questioned her, she gave herself away, said Panti. "That rotten old box with its cheap lock. Anyone could open it with a hairpin," she snarled.

"You see, she admits to knowing how to open it," Panti shouted to the police.

Panti wanted her arrested, and the police obliged by taking her to jail, where she spent four days before being released and freed of all charges. There was no evidence and no case, only accusations, he said, repeating what the police had told him.

"My heart is ripped out and still bleeding, Rosita. She might as well have stuck a knife through it," he said angrily, trembling from the gripping memory of such betrayal.

"Feel sorry for her, Don Elijio, because she will have to pay a high price for this sin," I said.

He nodded with such sadness in his eyes, as if her fate were the real tragedy. "Oh yes. She will die a painful, miserable death. Like one who is set out on a hilltop and tied to a tree, only to be eaten by the birds, piece by piece, until her hands rot off."

As if this tirade and confession had restored his fighting spirit, he said, almost defiantly, "I've been thinking, though. She really did nothing to me. I'm still here, neither more nor less. I have my dripping faucet of money going. It may never be a flowing creek, but it's a constant drip." With that, he folded his arms. "She hurt herself much more than she did me. God will punish her. I put revenge in his hands and commend her destiny to God."

As much as he longed to put this tragedy behind him, I could see he was still terribly upset. He kept muttering about feeling stupid and naive, as if he were a schoolboy who had learned nothing from ninety-two years of living. Could such a man be fooled by this evil woman? he asked, without needing a response.

"Don't blame yourself for being a loving man. She was the one who made a big mistake, not you," I kept trying to soothe him. "We are naive and trusting, but that's what makes us good doctors. It allows us to care deeply about other people's troubles. God made us that way, *papá,* and most of the time we do all right."

He was disturbed by the policeman's advice to put a lock on his door to keep out thieves. How could he do this—a lonely, old healer who opened his heart and home daily to strangers who wanted much more from him than his collected belongings?

But he did as the officer instructed, tacking up a metal bolt lock on a flimsy door that was made out of old scrap plank lumber. When we left the next morning to gather medicine, Panti bolted the lock and handed the key to Angel, who promised to guard it until we returned.

CHAPTER THIRTEEN

Cancer Herb   Hierba del Cancer
Acalypha arvensis

An abundant, small herb used to treat stubborn skin conditions,
infections, fungus, and wounds and drunk as a tea
for stomach upsets.
It may also be collected as one of the Nine Xiv formula for
herbal bathing, especially if used for skin ailments.

———————

It was nearly eight o'clock one evening the following October, and
Don Elijio and I had been sitting on the cement doorstep, watching the
sun go down behind the custard apple tree. San Antonio is on a rise, and
the sunsets there stretch for miles—long purple and magenta streaks
against darkening blue. It was one of those exquisite nights when the di-
urnal meets the nocturnal: the orange sun set on one side of the sky as
the silver moon rose on the other.

Don Elijio rose stiffly and announced it was time to go to sleep. It
had been a long and busy day with many patients. He began closing the
doors and windows as he usually did.

Down the path came an energetic group of women and children led
by Doña Juana, wife of Don Elijio's dear friend Don Antonio Cuc. She
was one of the village women who came to check on Don Elijio regularly, bring him treats and news.

Doña Juana was in her eighties, still trim with sharply etched Maya
features and silvery white hair. Like many Kekchi Maya women, she
wore a triple strand of colorful plastic beads tied in a knot at her throat,
and a cotton towel lay around her shoulders like a shawl.

She had fifteen children, upwards of eighty grandchildren, and
scores of great-grandchildren. She was herself an accomplished granny

healer and cared for her clan with simple home remedies gathered from her garden, nearby fields, and roadside paths. Like other granny healers, trained at their mothers' and grandmothers' knees, she sought Don Elijio's assistance when a family member didn't respond to her usual remedies.

We opened the door wide to let the flood of humanity through the gates of healing. They breezed in, smelling of soap and smoke and taking over the small room.

Don Elijio sat down at the consulting seat at his wooden table. Doña Juana took the patient's seat. Several young women accompanied her, each with several young children, and a teenage girl held a newborn baby wrapped in a light cotton blanket. One of the women was about seven months pregnant.

"I have brought two of my granddaughters and their children to see you," she announced.

Don Elijio looked annoyed. "Why do you not come in the daytime?" he asked. "The daytime is for healing, the nighttime is for sleeping. My useless eyes are worse at night. You're just lucky Rosita is here or you would all have to come back for your medicine tomorrow," he scolded in a not-too-stern voice.

"My granddaughters live at Mile 7 up the road and only got a late ride here," she explained. "We had to feed the children first, and tomorrow they must leave by dawn."

Doña Juana and Don Elijio conversed in Mayan for a while and then shifted to Spanish. As they spoke, the rest of us yawned, stretched, and chatted lightly about the weather, the moon cycle and planting season, and the crop of pineapples that year. It was clear her granddaughters couldn't follow the Mayan any more than I could. It looked to me like the great-grandchildren couldn't speak much Spanish either. They teased each other in Creole English.

"Now this daughter suffers from terrible headaches," began Doña Juana as she pulled the pregnant woman by the hand and sat her in a stool in front of Don Elijio. "I've given her teas, but it hasn't helped much yet. I thought you would have something stronger for her, little brother."

"Tell me, do these headaches come in the day or in the night?" inquired Don Elijio.

"Oh, my headaches always come just around two o'clock in the afternoon, when the day is the hottest," said the woman, whose name was Marina. "They can last for hours, even days."

"Uh huh," he said. "Daytime headaches need different treatment than nighttime headaches. Those that come in the day must be treated with a certain prayer and with cooled baths to the head and eyes. Nighttime headaches need hot baths and a different prayer."

"Ooooooooooh," responded Doña Juana, nodding her head approvingly toward Marina. "I told you he would know what to do."

"Rosita," said Don Elijio, not missing a beat. "Go to the other house and fill this bag with the Nine Xiv for Marina."

I used my flashlight to find the Xiv and filled the bag. The delicious aroma of the oils in the fresh leaves delighted and refreshed me. I wished I could sleep on a bed of freshly collected Xiv.

"Here you are, *maestro*," I said when I returned to the crowded room and handed Don Elijio the leaves.

"You are to boil a handful of these leaves for ten minutes, allow it to cool thoroughly, and then sit down in a chair with your head bent backward like this," he instructed. He mimicked the position he wanted her to take so that she would know how to do it. "Then, wash your head with the cool Xiv water and allow some to fall into your eyes. Cover your head with a thick towel and do not remove it for the rest of the day. Do this for three consecutive days."

Don Elijio reached out for her arm to feel her right pulse at the wrist, and as he often did, he motioned for me to feel the left.

"What do you feel?" he asked, peering at me intently.

"It seems rather slow and weak," I said. "Somewhat faint."

"Right," he said. "You have felt it correctly. The circulation is very poor. When I finish the prayer you will massage her neck and shoulders as you do mine. I know that will help her too."

When he finished the nine prayers, she shifted into a stool closer to the door where I massaged her neck muscles and corrected the alignment of her cervical vertebrae.

As I worked, the children sat wide-eyed and silent. They watched us like hawks, obviously drinking the scene in. Thus is traditional healing kept alive, I thought, by moments just like this, when matriarchs show their progeny how to get help from village healers.

No sooner had Don Elijio finished with Marina than Doña Juana pushed one of the smaller girls into the patient seat.

"Now, this little girl lives with the sniffles all the time," continued the matriarch. "Her father wants to get some pills for her, but I told him no, not to do that, as I have never seen that those pills work on people but instead make them tired and dizzy. We must try God's medicines first, I tell my children. If that doesn't work, then let's go see the doctor."

"Yes, sister, you are right," said Don Elijio. "Sometimes what the doctor can't cure, bush medicine can.

"Rosita, bring a small amount of Contribo vine for this child," he said. I released Marina's neck and shoulders. She looked disappointed. She had been enjoying the massage.

Back I went again into the dark hut, searching through the dozens of unmarked sacks for Contribo. I located it easily due to its strong, almost unpleasant, aroma.

Once again, I handed Don Elijio a ten-cent plastic bag stuffed with plants.

"This Contribo vine is to be soaked in a glass jar all day and then given to the child by the spoonful," he told the girl's mother. "Give her six tablespoons every day. Soon you will see that the phlegm worsens and seems to increase for a while, but that is good—it must all come out before she can be cured. Do you understand?"

"Yes, grandfather, I do," she answered using the Mayan word *nol* for grandfather. *Nol* was another familiar term of affection, like *tatito* and *viejito*.

I thought the family might be ready to leave, but Doña Juana shoved the other woman before Don Elijio.

"This daughter has a very serious problem with a fungus on her foot that is too stubborn for my treatments," she said forcefully. "Show him," she commanded the woman, who obeyed immediately.

This granddaughter, Josefina, was the mother of the newborn baby. She dutifully took off her plastic sandal and turned her left foot around so we could see the sole of her foot.

"Good grief!" I thought.

Don Elijio and I glanced at each other in disbelief. There were deep tunnels in her heel. Several fissures, the width of a pencil and an inch deep, cut into her flesh. The flesh itself was white, peeling, and cracked.

"What have you been using on it so far?" he asked Doña Juana.

"Like always, I bathed it with a hot, hot tea of Jackass Bitters," she said. "It's better than it was—if you can believe that—but it doesn't heal. What should I have done, *hermanito?*"

Don Elijio nodded and said, "The Jackass Bitters is good, but you should have bathed her first, then applied a dried powder mixture of Jackass Bitters and Cancer Herb. Those two are very powerful together. Do you know the Cancer Herb?"

"No *hermanito,* little brother, I don't," she said.

"Well, it's little, only a foot or so off the ground, with a little rounded leaf and a fluffy stick of a flower that looks like a tail," he told her.

She shook her head silently and said, "No, I haven't seen that one."

"No matter, you can come back tomorrow and Rosita will go with you to the old logging road and show it to you."

After he explained how she was to combine the leaves of both plants for the hot tea bath, he launched into a complicated prescription for the fungus. He told her to take equal parts of the leaves of both plants, toast them over the hearth until very dry, and pass them through a sieve to make a fine powder. Then she was to cover the fungus with castor oil and sprinkle the powder over the area, using a feather to make sure it went deep into the holes. He told Doña Juana to be sure that her grand-daughter kept her foot covered with a sock for the night. In the morning, she was to soak a sock in the hot tea and wear it all day long.

"That will do it for sure," he assured them.

The women nodded in respectful agreement.

"Is that all, then?" Panti asked, stretching out and glancing over at his hammock.

"Well, with your permission, I would like you to examine the big girl here. She has too much gas and indigestion all the time. She is constipated too."

Following them into the examining room with a lantern, I held it over Don Elijio's head as he dug deeply into her abdomen, giving her the classic *ciro* massage. The teenager squirmed under the pressure, but Don Elijio held fast.

The fingers of both hands dove into the soft flesh around her belly button, disappearing up to his second knuckles. Then he twisted his hands clockwise, pressing as he prayed.

"You have dry *ciro,* my child," he said as he turned her over on her stomach and massaged her back and legs with quick, squeezing movements.

"Rosita, go get her a mixture of Man Vine, Contribo, and Guaco. That will taste strong, child, but it will serve you well."

When I returned with the three-part formula, he reached into the bag and took out the amount she was to boil in three cups of water and drink before meals.

"Take no cold drinks, no acid foods, no beef, and no chili," he cautioned her. "Mark my words well, or you will be much worse off than you are now."

"Is that it?" he asked Doña Juana. "Can an old man go to bed now?"

"Well," she said, "one last question. This little boy is five and it is time for him to go to school, but he says he doesn't want to go.

"When his mother gets him up in the morning, he says he can't get up. He's too tired. He's even too tired to play sometimes. It isn't natural. I purged him of worms last month, but he is still not active and happy like the rest."

Don Elijio shrugged wearily and felt for the child's pulse. I reached out for the opposite frail wrist.

"Weak, thready, and too rapid," said my teacher. "Mark it well, for it is all too common."

Don Elijio picked up the lad's T-shirt and thumped his taut belly with his thumb and forefinger. Then he pulled down the boy's eyelid and asked me to tell him what I saw.

"Is it red, pink, yellow, or white?" he asked me.

"It's pale and on the edge of yellow," I reported.

"Yes, just what I thought. The child is anemic and has a tapeworm, probably a large one."

"Uh oh," commented his mother. "Roberto is very stubborn about taking medicine. He cries and fights, even spits it out when you're not looking. I'm afraid he won't swallow anything bitter or strong tasting, *nol.*"

"Don't worry about that, I know what to do," said the old man. "You will take the roots I give you and boil them in sugar water for half a day until the brew is as thick as syrup. Give him this syrup six times daily by the spoonful, and on the third day he should fast. On the fourth day, give him a strong dose of castor oil. Soon, *mamasita,* your son will be strong and happy again, I promise. Have faith, God will help us all if we but ask.

"I know all these things to be true, because I have tried and tested them," he continued. "I've been doing this work now for forty years, and I know a few things."

The children were getting restless and starting to fuss. They started to get up to leave, when Josefina paused and timidly said, "Please *tatito*, I have one more question.

"It seems I don't have enough breast milk to satisfy the baby," she said.

"Doña Juana, you know the wild Poinsettia plant," he said. "Doesn't it grow right in front of your house?"

"Yes, yes. That one I know for sure, but my grandmother told me never to use it, because it is poison."

"True, very true—smart woman," he said approvingly. "You were wise to obey her counsel. But in this case, it is safe. I want you to pick nine of the little plants and braid the branches into a necklace for her to wear all day. For nine days you should make a fresh necklace every morning. Also boil a pot of the herb and wash the breasts with the warm water before each feeding. Dry the breasts before giving them to the baby. Then, ha, there will be quarts of milk."

Doña Juana looked at Josefina, who was now nursing the infant at her breast, and said authoritatively, "The old man has very sore and weak eyes. I think a little bit of mother's milk dropped in them would be helpful. Are you willing, child, to give him some?"

Josefina giggled and shifted uncomfortably, turning to the wall to blush. She covered her mouth with her free hand and whispered, "Sí."

Then there ensued a great debate on how best to squirt the milk into Panti's eyes.

I have to see this, I thought.

Finally it was decided that he would remain in his seat and she would stand up and point her nipple at his eye. It was my job to hold his eyes open to allow each milky projectile to find its target.

Amidst a great babble of giggles and shifting of feet, Don Elijio gladly gave himself over to Doña Juana's prescription. How easily and quickly they switched roles, I thought, each helping the other with their own brand of expertise. After fifteen children and eighty grandchildren,

I guessed Doña Juana knew everything there was to know about the health benefits of mother's milk.

He sat expectantly on his little wooden stool in the corner of his poorly lit cement hut, beneath his calendar sporting a half-naked woman in a bathing suit who was guzzling a bottle of Coca-Cola as an ocean wave broke over her.

Doña Juana held the lantern overhead. Josefina positioned herself directly in front of Don Elijio's face. She blushed and giggled. He braced himself and said something in Mayan that made the women laugh, and then the streams of milk shot forth one by one into his waiting eyes. He blinked, giggled, and blinked again. The children squealed in delight.

"More," demanded Doña Juana, and Josefina obeyed with two more squeezes that also found their mark. "Hmmmm, feels very warm and soothing," commented Doña Juana's patient.

Milk streamed down Don Elijio's face as he turned to me with an impish smile. I wiped the droplets off his cheeks.

At last they all left with a hundred words of gratitude. Each woman paid him what she could and called on God to grant him a long life. Don Elijio had no set fee for his services, but most patients paid between five and fifty dollars. Some could only pay with a thank-you, which Don Elijio said was more valuable than money because it was a direct blessing from God.

Doña Juana and I made a date to go out and look for the Cancer Herb the next morning right after sunbreak.

Later, as we swung in our hammocks discussing the day through the cotton curtain, he warned me not to pick the Cancer Herb until sun had risen and the dew had dried off the leaves.

"The plant's healing power is still down in the roots until the sun calls it up to the stem and leaves," he explained.

"Then would you better collect roots in the early morning rather than in the full sun of the day?" I reasoned.

"Yes, girl, that is why we leave home so early to dig the roots and only collect the Xiv on the way home after the sun has risen high in the sky and dried off all the dew. To everything there is logic. All of this is in my head. I never went to school, can't even sign my name, but still my head is full."

CHAPTER FOURTEEN

Chicoloro   Strychnos panamensis

Considered one of the primary medicines in Maya healing.
The woody vine is chopped and boiled as a tea to be
drunk for gastric conditions, uterine problems, poisoning, and
constipation. The active principle, strychnine, is toxic when
taken in excess. The cross pattern on the branches is considered a
warning that the plant is medicinal but toxic and
must be consumed with caution.

A beautiful Maya woman from Corozal District up north showed up at
the clinic one day because she wished to stop a seemingly endless pro-
gression of children.

The woman, Berta, told us she was thirty-eight and the mother of
fourteen children and the grandmother of six. "Too many," she told us
with a lovely, gold-toothed smile. Despite the work of bearing and rais-
ing such a brood, her dark skin radiated health and her ebony eyes
glowed brightly beneath a forehead that sloped back into a thick, black
braid.

She was one of those strong Central American women whom I so
much admired: contented, conversant, in charge. She conjured up an
image of the wise woman at her helm of the family ship. I could picture
her hand-scrubbing clothes for twenty people, grinding corn, working
in the fields, cooking over an open hearth, nursing babies, attending to
her husband—all with grace and laughter.

What Berta wanted was a natural birth control method, she explained to Don Elijio. It was possible, at her age, that she could still bear
five or six more children.

"I love my husband and family very much," she explained. "But
fourteen children is enough, I tell him. He doesn't want me to use any

pills, so he told me to come to see you to seek your help for something natural."

Don Elijio laughed and grinned his usual jolly, toothless grin. "I fix those who want and those who don't want," he said simply, repeating the line he said so often. He was very comfortable with discussing birth control, as women made up a large portion of his patient load and troubled and ailing women came from all over Central America to seek out his help and advice.

"Yes, *mamasita,* I know what you need," said Don Elijio. "It's the *Ki Bix* or Cow's Hoof Vine." He turned to me and said, "This plant will be very important for you to learn, Rosita. I use it for birth control and dysentery. It's safe and sure." I had long wondered what plants Don Elijio used for birth control. The birth control pill developed in the 1950s had been a gift of the Mexican Nahuatl women, who had long used the young Wild Yam root as an effective birth control agent. They had shared their knowledge with biochemist Russell Marker, who eventually brought it to the attention of research scientists. The Wild Yam root contains diosgenin, a steroid that mimics pregnancy hormones, tricking the body into believing it's already pregnant. Wild Yam grew abundantly on Ix Chel Farm.

I was anxious to see the *Ki Bix*. In order to work it had to be freshly harvested, said Don Elijio. But since it was already late afternoon—too late to walk to our rainforest *farmacia*—Don Elijio told Berta she'd have to spend the night. He was obviously pleased at the prospect of enjoying her delightful company for the rest of the day. He showed her the bed that was really a wooden door and gave her a lantern he had made himself—a glass jar filled with kerosene and a little strip of rag.

The silver mists still hung over the village when Don Elijio and I left for the high forest at dawn. *Ki Bix,* or Cow's Hoof, was on the top of our list for plants to collect as we climbed upward toward the Mountain Pine Ridge. Empty sacks, picks, and shovels in our hands, machetes in their leather scabbards hanging at our sides, we followed a new path toward the crest of a hill where he had last seen the woody vine.

It turned out to be more than a ninety-minute walk to that hill. Usually we skirted around the foothills of the Maya Mountain range and the Mountain Pine Ridge. Today we headed straight up, with the razor-back hills below us.

As we climbed, the forest changed from graceful palms to stately pines. Alone in the quiet coolness, we didn't talk so as not to disturb his silent search. He murmured a prayer. The only words I recognized were *Ix Chel* and the Mayan word for woman, *Colay.*

"Ahh, here you are," sang out Don Elijio, as if greeting an old friend. I was impressed as always by the old man's memory. His mind was a map of botanical treasures. It was as he said, all in his head. I thought, it was all in his heart.

He showed me how the female vine or *Ix Ki Bix* (*Ix* means "female" in Mayan) grew right beside the male plant. It was the female, he said, that he used for birth control. The male was used to stop hemorrhaging and dysentery.

The male was an enormous, if spindly, rough-barked dark vine that stretched precariously many hundreds of feet into the dappled sunlight of the rainforest canopy. It looped itself around branches of the towering trees. Three feet away from the large male trunk was the female. Her vine, smooth barked and bearing three-inch thorns, gracefully loped around the same branches as the male, as if in pursuit. Twenty feet in the air above us the male and female *Ki Bix* entwined in an embrace.

"*Amantes de la eternidad,*" giggled Don Elijio. Eternal lovers.

Don Elijio scratched the bark of the male vine and showed me the white inner bark. He told me to scratch the bark of the female with my machete. I did and uncovered a mahogany-colored inner bark. The vine was red and layered throughout, resembling the female uterine membrane.

I couldn't take my eyes off the redness of the vine. I was always amazed by nature's way of letting us know what a plant might be used for by matching the color or shape with the complaint. I had noticed that this relationship between color and use—the Doctrine of Signatures—seemed most evident when it came to plants connected to women's needs.

Many Maya medicinal plants used for women's ailments were reddish of tint, and often the female leaf was broader than its male counterpart, as was true of *Ki Bix.*

"See how this leaf is split and looks just like a pair of trousers," said Don Elijio.

I couldn't resist a joke and said, "The sign here is clear. Keep your pants on."

We laughed until our eyes teared. Then we got back to work. We cut a twelve-foot length of the female vine, chopped it into one-foot sections, and stuffed them into my sack. We then cut some of the male vine and, as was our wont, went on to collect the leaves of the *Ki Bix* as the first of the day's Nine Xiv.

Many hours later, we returned to the clinic, laden with our usual cargo. Immediately after lunch, Don Elijio had me chop the *Ki Bix* vine for Berta.

He instructed her to boil a handful of the vine in three cups of water for ten minutes and to drink one cup three times daily during menstruation for nine months consecutively. This would prevent pregnancy for the rest of her life, he assured her.

"You could sleep with your man six times a day in the bed, in the bath, in the car, anywhere you like, and as many times as you like, and nothing will take. Only groans of pleasure. Nothing else. Don't worry. These things are no mystery to me. It's all up here in my head. Right here."

Berta laughed vigorously, paid Don Elijio five dollars, and, clasping her bag of *Ki Bix* against her breast, found a ride going toward town in a truck overloaded with mahogany logs.

That evening, after all the patients had left, I asked him how the *Ki Bix* worked. "*Se secca el cuajo,*" he explained. "It dries the membrane." This prevents implantation of the fertilized egg.

Treatment for one month would last for five months; two months of use provided ten months of protection, he said. Women who took it for nine months would become permanently sterile.

"No woman has ever come back to say she was pregnant," Don Elijio told me.

I asked him if he was sure. He answered, with his usual flippant humor, "Hmmmm, Rosita, no babies named after me!"

I saw over time, however, that the *Ki Bix* or Cow's Hoof didn't, in practice, seem to be 100 percent effective; implantations did occur. For instance, the woman who lived in Cristo Rey Village two miles down river from our farm requested the vine, then became pregnant in the third month of the five-month protection period. An American woman from Wisconsin, who had heard about Don Elijio through friends and was thrilled to have some form of natural birth control, wrote me that she also got pregnant in the third month. Still, many women reported that they didn't conceive and swore by the thorny vine.

When I mentioned that to Don Elijio, he said, "Sometimes they don't take it the way I tell them to. And sometimes, she secretly wants to be pregnant."

Ki Bix was only one of the many plants that Don Elijio favored in the treatment of women's ailments. Contribo vine was excellent for menstrual complaints, especially if complicated by gastric problems. The bark of the Copalchi Tree was used for diverse symptoms such as painful periods, infertility, menstrual migraines, and hormonal imbalances. Most often he made a mixture containing Contribo, Copalchi, Man Vine, and *Zorillo*. He called this *Sacca Todo,* pull out everything, in reference to its marvelous eliminative powers and ability to cleanse the uterine membrane of incompletely flushed menstrual fluids. Incompletely flushed fluids, with consequent hardening of the membrane, was one of the main causes of painful periods, he said.

While plants were an essential part of his treatment of women's problems, his philosophy of women's health centered on the proper position of the uterus. Don Elijio believed that displacement of the uterus was what caused the multitude of women's complaints. He said that 90 percent of modern women have a prolapsed uterus that lies askew to one side or the other, tipped backward or forward, or too close to the pelvic floor.

"The womb is the woman's center," he said. "Her very being and essence are in this organ. If the uterus is not sitting where it should be, nothing is right for her; she will have late periods, early periods, clotted blood, dark blood, painful periods, no babies, weakness, headaches, backaches, nervousness, and all manners of ailments."

He treated uterine displacement with vigorous external massage, which sometimes made the women cry out in pain. This was followed by *Sacca Todo* taken as a tea—three times daily, ten days before menstruation—as well as vaginal steam baths with Xiv.

I watched him one day in action. Lola, a mother of three young children, was complaining of extremely painful periods that made her go to bed for three days at a time. She had had three hospital births in four years, and each one had been difficult.

"It was very hard for me to have my babies, *tatito*," Lola cried, "and it took me a long time, a very long time, to recover afterward. What is wrong with me? The doctors and nurses say there is nothing wrong, but I know something isn't right. I come with faith to see you."

"Come, I must examine you," answered Don Elijio, motioning to her and to me to follow him into the examining room.

Lola lay down on the mattress and raised up her skirt, exposing a soft brown belly with many stretch marks. He began to examine her abdomen just above the line of pubic hair. Within a minute he exclaimed confidently, "Aha! There's your sickness. The womb is out of position. It's too low and lying on the right side. You have headaches, numbness in your right foot, weakness in the legs, and constipation. Your periods are early, then late. You see clots of blood, dark blood. You feel tired often. Is all this true?"

"Why yes, *tatito*, it is all just as you say!" Lola shot me a look of shock and wonder. "What should I do?"

"It is what I will do, not you," he answered as he went to work.

Deep, deep into her flesh he dug his gnarled, expert hands. Lola groaned and stiffened in discomfort. She covered her face with her hands and turned away.

"Come here, Rosita, feel this," he said.

He manipulated her abdomen to expose under his fingers a soft, bulbous object off to the right side just above the pelvic bone. Copying his hand movements, I too could feel the uterus. I saw what he meant: it did seem too low and too far to the right.

"Now watch," said Don Elijio, smiling. He leaned over Lola and pulled upward with his hands, dragged her skin to a center position, and then jiggled the flesh vigorously. He repeated this again and again, then slowly pushed from the right side of her abdomen to the center, and from the hollow above the pelvic bone upward. As he kneaded and massaged, he whispered prayers.

I could see the uterus responding and moving. "Now feel this," he said. I put my hands on the woman's pelvis. Where the uterus had been just minutes before was a soft hollow. The organ had moved to its proper place quickly and, this time, almost painlessly under his deft care.

He showed Lola how to use a *faja,* a band that is tied around the pelvis to help hold and heal the uterine ligaments. As she sat up and straightened her clothes, he cautioned her sternly about going barefoot on cold floors, especially cement floors. He told her not to take cold drinks, then warned her against having sexual relations with her husband when she had her period. The latter, he explained, was the greatest abomination ever perpetrated on women.

"The semen mixes with menstrual fluids and causes large clots to adhere to the uterine wall," he told her. "This leads to ulcerations and tumors."

He motioned for us to follow him back into the other room, where he rummaged through his plant bags and prepared the *Sacca Todo* mixture.

"Drink the medicine for ten days before menstruation for the next three months," he said, "then come back when the medicine is finished." He told her to expect some thick, dark fluids to pass. "Don't be alarmed," he assured her. "That is your sickness coming out."

"Better an empty apartment than a bad tenant," I joked, using one of my mother's favorite lines. Both Don Elijio and Lola looked at me and giggled.

After she left, I asked him why it is that so many women have displaced uteri.

"Modern life," he answered laconically, "carrying heavy loads too soon after childbirth. Midwives, doctors, and nurses who don't put belly bands on the woman after delivery to ensure the uterus is returned to its rightful place. That's bad care.

"Also those horrid, ugly shoes with the sticks in the back," he bemoaned. "And walking barefoot on cold floors and wet grass, especially in the early morning hours."

He said nervousness and anxiety in modern women also exacerbated uterine weakness. When a woman's muscles were tense, the blood supply to the uterus decreased, thereby setting the stage for problems to develop.

Don Elijio was famous for his ability to correct uterine displacement. Once a taxi full of young and middle-aged women arrived from San Ignacio for the express purpose of having their uteri replaced. Each recounted a string of familiar symptoms, and, one by one, he led them into the examination room.

One woman in her sixties had a uterus that was particularly far afield. It was lying nearly below the inguinal ligament, just above the thigh. He instructed her to lie on her stomach as he skillfully executed a sophisticated chiropracticlike technique to her sacrum.

He pressed down on the small of her back as he brought both feet towards her buttocks, simultaneously executing the forward motion of one hand with the backward pull of the legs. He did this, he told me, in chronic cases of longstanding displacement to strengthen the ligaments that hold the uterus to the sacrum.

Her daughter, a grandmother in her forties, came in afterward. She complained of a bothersome yeast infection that did not yield to usual medical treatment. "They give me the medicine, I take it, the itching and burning goes away for a while and then comes right back," she lamented.

This time he sat down on a stool and said, "You do it, Rosita. Tell me where you find it."

Not too confidently, I began to examine her. I was amazed at how easy it was to determine the position of her uterus.

"I find it very low but in the center," I told him.

Don Elijio got up and felt her belly. He nodded and smiled. "This is easy for her," he told the woman proudly. "She knows the body well and is already a doctor. And a woman."

"When we put your uterus where it should be, the itching will stop," he said to the woman. The weight of the uterus on the vaginal tissues prevents proper flow of blood, lymph, and nerve currents, thus allowing the yeast to thrive. When the uterus is properly placed, the elements of the blood will adjust the pH of the vaginal wall, thereby making an unfavorable environment for the yeast.

Don Elijio returned her uterus to its proper place. "See here," he said, pointing to the pelvic bone. "The width of two fingers above this bone is where a woman's uterus should sit. No more and no less. And always be sure it is in the center. "

Because of his knowledge, Don Elijio had become legendary among midwives. In Belize, many women still rely on lay midwives for childbirth. In a country of remote villages with few ambulances and hospitals, midwives are accepted and respected members of the basic primary health care team. The country has an excellent system of training home birth attendants, in which women who want to become midwives are trained by other women.

San Antonio, like all other villages, had two or three women trained as midwives who took care of the vast majority of home births. They came to Don Elijio only when a patient was in serious trouble. Don Elijio had never lost a patient in childbirth. He was expert at coaxing intransigent babies to come out or to turn. He dealt with breech births with a series of manipulations and special prayers.

"I've never had to send a woman to the hospital," he said. "My prayers and massage have helped every time. But the doctors are in such a hurry nowadays. Twelve hours of labor pass and they sharpen their knives."

Once on a cold, rainy, winter night in January, a midwife came to his house and roused the old man out of his sleep.

"Get up, old man, I need you," she said as she frantically knocked on his door. Don Elijio grabbed the plastic purse he used as his doctor's bag and followed her back to the patient's house. A baby—a girl—was already born and lay swaddled in the arms of a frightened ten-year-old boy. Their mother lay prone on a mattress of cloth rags, her uterus protruding from the birth canal. There it lay, he explained, like a pink balloon between her legs.

"I scolded the midwife for making the woman push too long and too hard," he said. He called for someone to make a fire in the middle of the dirt floor. He took out a bottle of olive oil and a small clay bowl from the purse, then warmed the oil in the bowl over the coals. He poured the oil over his hands and rubbed some onto the dislodged uterus.

Gently and slowly, whispering his Mayan prayers to Ix Chel, Goddess of childbirth, he gradually set the uterus back inside the pelvic cavity. "I heard it pop as it went back into position," he said.

He asked for clean sterile cloths, which he pushed into her vagina to hold the uterus in place. Then he tied the *faja* around her pelvis to hold in her overstretched ligaments and gave her the baby to nurse, knowing that nipple stimulation contracts the uterus. An hour later, he removed the cloths and allowed the postpartum fluids to flow freely.

The woman recovered completely. "That was my twenty-seventh godchild," he said proudly. "Her name is Gomercinda. They call her Chinda."

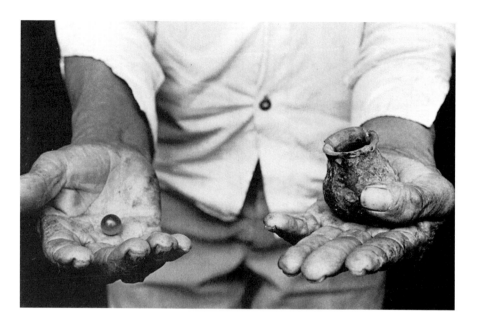

Don Elijio holding his sastun *(photo by Carol Becker)*

Rosita with Don Elijio on the trail *(above)*
Rosita on expedition *(right)*
Don Elijio with patients *(opposite, top)*
Don Elijio, Rosita, and Mike Balick
(opposite, bottom)
(photos by Michael Balick)

Jackass Bitters *(above, left)*
Contribo *(below, left)*
(photos by Michael Balick)

Gumbolimbo *(above, right)*
Cow's Hoof Vine *(below, right)*

Polly Red Head *(above)*
Wild Yams *(below)*
(photos by Michael Balick)

Don Elijio gathering herbs on trail *(above)*
Inside Don Elijio's house with herbs *(opposite, top)*
Rosita with Don Elijio chopping medicine *(opposite, bottom)*
(photos by Michael Balick)

Don Elijio on trail to village *(photo by Carol Becker)*

CHAPTER FIFTEEN

Balsam Tree Balsamo Na Ba
Myroxylon balsamum

The tree has a highly resinous and aromatic bark, which is used
to treat both physical and spiritual diseases. A small square of
the bark forms part of the Maya amulets or *protección,* carried
to ward off envy, the evil eye, evil spirits, and black magic. The
boiled bark is a primary remedy for conditions of the
urinary tract and the liver.

The film *Mosquito Coast* was being shot in and around Belize. It's a story
about an obsessed man who brought his family to Central America to
live in a jungle, where hardships eventually pitted them against each
other and revealed the father's coldness and selfish obsession with per-
fection.

One of the film crew's favorite resting spots was Chaa Creek, by
now the most famous of all the jungle resorts in Belize and known for
its primitive elegance. Chaa Creek had a lush, tropical landscape with
thatch roof bungalows and a stunning open-air bar overlooking the
river.

Lucy and Mick introduced Greg and me to the film's editor,
Thomm Noble, who we learned had won an Academy Award for his
work on the film *Witness.* We became friends. After a few days, Thomm
and other crew members invited Lucy and me to accompany them on a
trip to Tikal, the famous ancient Maya city ninety miles west of us in
the Petén region of Guatemala.

We crossed the Belize-Guatemala border at an outpost about four
miles west of Ix Chel Farm and headed west on the narrow dirt road
through the jungle. It was a hot, bumpy ride. We sat on wooden

benches in the back of the truck bed with dust swirling through the air and dirt flying into our faces.

The city-state of Tikal flourished in the Classic Maya era and is known for erect, graceful pyramids and its dominant position in the Maya lowlands. Until the early part of the twentieth century, the ruins of the city were covered by the jungle. When the city was rediscovered, it was excavated and reconstructed by archaeologists from the University of Pennsylvania. That work is now continuing under the Guatemalan government. Today it is a vast park in the jungle in a sparsely populated area of northern Guatemala, once the favorite haunt of *chicleros* like Don Elijio.

It is hard not to be struck by the mystery of the Maya when entering the now-silent, stunning city. What happened to the Maya civilization? Why did the Maya suddenly abandon cities like Tikal over what is today Mexico, Belize, Guatemala, Honduras, and El Salvador around A.D. 900? What were the ancient Maya like?

Lucy and I and our Hollywood visitors wandered around like children at a county fair, stopping to gaze at the temples, stelae, and picture carvings of the ancient Maya.

I couldn't help think about Panti's superstitions about Tikal. It was here that he and Jerónimo had studied plants, and he had warned me that the Maya Spirits who haunt the place could be dangerous. He would never set foot in the excavated and reconstructed Tikal. But I had been here many times and had never felt uncomfortable.

Thomm and I climbed to the top of Temple IV together and sat there for a long time enjoying the sweeping panorama of the rainforest below. The ghostly gray temples, once covered in natural red dyes, shimmered in the heat. The same dry winds that nearly destroyed the farm cooled us.

Thomm pulled out several lovely, large crystals from his pouch and spread them out in the sun. We watched the light play among the planes of the crystals as an eagle flew back and forth over our heads, almost daring us to reach out and touch him. We sat quietly, content to be surrounded by the sky and the howling winds.

The others called from below, and we descended the ragged, uneven steps. We both felt happy and peaceful when we got to the ground.

We arrived back in Belize late that night. I had a busy day ahead of me on the farm with patients scheduled back to back. Shortly after noon, Lucy and Mick's young son, Piers, came skipping over clutching a note. It was from the film crew, informing me that Thomm had taken ill with *turistas* during the night. Could I help? the note pleaded.

In between patients, I prepared some of our Traveler's Tonic, a brew we concocted of Jackass Bitters and wine to treat and prevent parasites and amoebas. I sent Piers back with a large bottle and a note with instructions.

Later in the day, one of the crew, a beautiful Eurasian woman, came over to say that Mr. Noble was grateful for the remedy, which had cured him, enabling them to travel again.

I was relieved to know he had recovered and pleased that the tonic had cured again. The woman left without offering to pay for the tonic, and a part of me was annoyed. After a few seconds, my annoyance passed. I let it go, telling myself: May God repay. I went back to my patients and didn't give the matter a second thought.

At dusk, Piers was back again. I'm always glad to see him and walked outside the kitchen to greet him.

"Mr. Thomm sent me over to give this to you," he said in his sweet, baby boy voice.

He opened his chubby, pink palm and plopped a gorgeous, oblong-shaped crystal into my hands. It was the loveliest of the crystals I had seen atop Temple IV glittering in the sun. Rainbow-colored areas of light shimmered within. It was huge—four inches long and three inches wide, coming to a graceful point at the top. Its size and weight felt perfect in my hand.

I wanted to run over and thank Thomm, but Piers told me that the crew had left and that Thomm had instructed him not to deliver the crystal until after they had driven away.

The next time I went to San Antonio, I brought the crystal with me to show Don Elijio. The moment he saw it, he jumped out of his seat, threw up his hands, and shouted, "Sastun! Sastun! They've sent you a sastun. How wonderful, my daughter, this is the sign I've been waiting for. Ix Chel sent this to you. Now I can teach you everything, everything. You can work in the spiritual as I do."

"This is a sastun?" I sputtered. I flipped the heavy crystal over and over in my hand, waiting for its mystic abilities to become obvious.

"Yes, girl. This is not just a simple gift. This came right from the Maya Spirits. They used your friend as a carrier."

He opened both his hands and gestured for me to place the crystal in his palms. He pushed back his Pepsi cap and stared intently into the stone. "Uh huh. See here." He pointed to a corner of the crystal and showed me a rainbow-colored cross.

I really couldn't believe what Panti had said. The Maya Spirits—the nine celestial beings at the core of the Maya culture—had sent *me* a sastun?

My sastun didn't look anything like his small marble. I didn't see how I'd be able to twirl this chunk of crystal the way he twirled his.

"Can I use this to get answers the way you do?" I asked.

"Well, I'm not sure," he said. "The sastun can take many forms. I once had a sastun that I used only to see if a patient could be cured. It was thin as a pencil with a dot in the middle that would stretch from end to end when a patient who could not be cured held it in his hand. I once saw a blue sastun on a necklace."

"How do you know a stone is a sastun?"

"I know by looking at it," he said. "It has light that sparkles when it moves and you can peer into it as a mirror. You may find dots, lines, crosses, Virgins, and rainbows that give you answers. If you have the lamp to see it."

"What do you mean by 'lamp'?" I then asked.

"It's something up here that's a *don*," he said tapping his forehead. "A gift."

Clearly Don Elijio had the gift. He regularly used his sastun to determine if a patient's illness was rooted in natural causes or from *daño*, meaning brought on by spiritual forces. It was his sastun that carried requests and prayers beyond the gossamer veil to the Maya Spirits. In using it, he was carrying on an ancient tradition. Archaeologists had found sastuns in the burials of Maya shamans in the abandoned ancient cities.

My image of a sastun was that of a supernatural hot line to the Spirits. Don Elijio described it as a Stone of Light, Mirror of the Ages, Light of the Ages, and Stone of the Ages. He also called it a "plaything of the Maya Spirits." He said they could be heard at night throwing a sastun back and forth to each other across rushing rivers, which fail to drown out the whirling sound of its fanciful flight.

The Spirits chose to whom to send the sastun. Some people prayed for a sastun all their lives but never received one. "Others don't know what it is when it rolls by their feet while they're playing in the sand," Don Elijio said. "Often it falls from the sky in front of someone. But many fear it and try to throw it away. This is useless, as it will only come back."

They fear it, as Jerónimo had, because of its drawing power. "When the Spirits sent me my sastun, the people came from very far because it calls them. Like me, it loves to work," Panti told me happily.

Panti wanted to consult his sastun to verify that the Spirits had indeed sent me this gift. He lifted the little clay jar out of the rusty Ovaltine can where he stored his sastun.

He removed the washcloth that he kept stuffed into the neck of the jar to prevent the sastun from rolling out. Then he blew into the jar before spilling out his round, translucent marble into his hand. It sparkled, having just received its weekly bath, in rum, to cleanse it of the many questions he had asked it during the week.

He blew on it three more times, then placed it back into the jar. My crystal sat on the tabletop, while he twirled the jar around it, uttering a Mayan chant. I heard the word *sastun* and my name repeated several times.

He motioned for me to open my right palm to receive his sastun, just as I had seen him do with hundreds of patients. I took it in a loose fist and shook it like a pair of hot dice at a Las Vegas casino.

As I shook, I began to get nervous. Could Panti be wrong about this? If it is a sastun, why was it sent to me through such an ambiguous path? If it is a magical instrument, what am I to do with it?

Would the sastun draw patients to me? Did I want people seeking me out as they did him? How would I handle the types of cases that required a sastun?

I felt like turning and running away.

"Come to the doorway, where I can see," said Panti, so excited he didn't notice my confusion. "My eyes are terrible this week," he continued. "I think I'll be blind before long, then who will there be to pull me around?" he asked, not really waiting for a response.

I opened my trembling palm and he moved it back and forth until his sastun danced about on my clammy skin.

"It is done, my child. This is a sastun," he said firmly. "The Maya Spirits have sent it to you. They have accepted you."

I slumped onto a wooden stool and blankly watched him chant over my crystal in Mayan, moving his clay *jarrito* about it in some sort of merry, consecrating dance.

My thoughts tumbled over each other in an unruly frenzy. I was thrilled. I was frightened. I believed. I didn't believe. Was this foolishness or guidance?

Closing my eyes for a moment, I leaned against the hot cement wall, drawing in a slow, deep breath and praying quietly, muttering holy words as fast as I could think them. All at once, a calm slowly draped over me like a protective cape, and I began to relax, letting my shoulders gently slide down the concrete. The words "Have faith. . . . Thy will be done," seeped into my mind.

Jarring me out of my paralysis, I heard Panti's raspy voice. "I'm enchanting this stone for you, Rosita, asking the Spirits to show you in a dream how you are to use it, read it, and care for it. A sastun demands much attention, you know."

"I know, Don Elijio, that's what I'm afraid of," I mumbled.

"Afraid? Are you really, my daughter? Why?"

"Because I don't think I can do what you do, *papá.*"

He waved away my objections. "Don't think about all that. Day by day. Step by step, Rosita. You've already learned so much. I always tell people how quickly you learn. What would happen to me and my art were it not for you, child? Just think of that."

I had thought of that, and the concept now seemed overwhelming. "The Spirits frighten me a little. Should I be afraid of them?"

"No, no. They are good friends, very gentle, wise, and loving," he told me. "Have no fear, I know them well. They can be mischievous, but never intentionally harmful."

Then the old man said, "You'll be meeting the Spirits tonight."

CHAPTER  SIXTEEN

Allspice   *Pimienta Gorda*   *Pimenta dioica*

The berries and leaves of this common tropical tree are used as
a household tea to correct indigestion and childhood colic and to
warm the body during cold weather. Women make a sitz bath of
the leaves boiled in water to relieve menstrual cramps. The berry
is a popular spice added to many local Belize dishes. The crushed
berry is part of the ancient Maya embalming formula.

The Maya Spirits use dreams, Don Elijio told me, to communicate with
people they have chosen as H'mens.

As a warning, so I wouldn't be frightened by their arrival, he described the sensations I would experience—a sort of tugging and
pulling just before the dream begins. "Sometimes it even feels like little
gremlins are crawling all over you with furry bodies and bony hands.
But don't panic, whatever you do, because then the Spirits will leave."

"What? Gremlins! Furry bodies! Bony fingers! In my sleep? Don
Elijio, do they do that to you?" I could hear hysteria creeping into my
voice.

"Yes, child, almost every night. I'm used to them now. Sometimes I
tell them to go away and let an old man sleep. They come to look for
companionship and to let me know that a dream is about to begin."

He instructed me to say an Our Father if I felt afraid. It would protect me from the evil spirits, who also try to slip into dreams. "Now,
they are frightening and evil, but that won't happen if you do as I say."

"What if I faint or scream out?"

"You won't. Just have faith. Have faith. I'll teach you a new prayer
to empower your sastun. You will gaze into the crystal and say this

prayer nine times every Friday ever after. While you pray, make the sign of the cross on both sides of the sastun, dipping your finger into a bit of rum.

"Sastun, sastun," he chanted. "With your great power I ask that you tell me all I want to know. Teach me to understand the signs and visit me in all my dreams to give me the answers I seek. I have faith with all my heart that this sastun will answer my prayers. God the Father, God the Son, and God the Holy Spirit."

Before he died, Jerónimo had warned Panti that he would receive a dream visit from the Spirits after he received his sastun. Jerónimo had told him that dreams were the lines of communication the Spirits preferred and that every dream would leave him feeling like he would live forever. Jerónimo had also told him that he would never forget any instructions revealed in a dream.

Panti's dream occurred the first night he slept with his sastun. After drifting off to sleep, he felt his hammock was being pulled from side to side by a pair of forceful and determined hands. He was frightened at first, but within an instant the dream vision began and his anxiety gave way to anticipation.

Suddenly, standing before him was an ancient Maya, striking and confident in a white tunic that barely brushed the tops of his knees. The robe was gathered at the waist with a belt dripping with regal strips of animal skins, glorious remnants of the revered jaguar. The Maya wore *macasinas,* a simple foot covering of rubber soles and straps coming up between the toes that fastened around the ankle. In the middle of his bare chest dangled a sparkling jade pendant, and crowning his head was a feathery headdress of large, brightly colored plumes.

Another Maya, wearing a simple cotton tunic, stood by him silently.

The first Maya clutched his carved staff, which resembled a writhing, menacing snake, and said, "Elijio Panti, we see you are working hard and we are pleased. We send you now this sastun to help your work. Use it to help people but do no harm with it. Wash it every Friday and use the following chant when you wish to ask it a question."

The ancient Maya repeated the chant that enabled the sastun to be blessed and then taught Panti how to read the bubbles that would form inside the marble as answers to his probes. Before departing, the Maya told him that he and other Spirits would visit him often in the future.

As soon as he awoke the next morning, Panti realized that the ancient Maya was one of the Maya Spirits. Just as Jerónimo had promised, he recollected all the details of the dream, and he was filled with a sense of wonderment and well-being.

That morning Panti held his sastun, turning it, washing it, and rolling it about inside an old cup, making it twirl the way the Maya had showed him in the dream.

He was both excited and bewildered, but in his heart he knew he was worthy of the gift. He had not prayed for a sastun to bring him fame or great riches or to harm someone through its power to enchant. He had desired a sastun to be able to cure more of his patients' ills.

This was not the first time that I'd heard Panti speak of his dream visions. Through dreams, the Spirits delivered valuable information to him. Whenever he was confused about a patient's illness and didn't know how to treat it, he consulted his sastun, asking the Spirits for their help. They often answered him through dreams, he said, showing him which plant to use, where to find it, how to use it, and what prayers to say in accompaniment.

"The next day I would grab my bag and my machete and go off to the mountains to hunt for that specific plant. And I always found it right where they said to look for it."

This talk of gremlins, dreams, and ancient Mayas was making me more anxious about what lay ahead for me. I didn't really know what to make of Don Elijio's certainty that I would have a dream vision that night. By now I had tremendous faith in him and was sure that he had never lied to me. He seemed calm, so certain that the Spirits would visit and that I wouldn't die of fright. I was not so sure, but I decided to accept what he said and have faith in his wisdom.

By now it was night. We ate a light meal of beans and tortillas. San Antonio reverberated with sounds of the night. Couples carrying boom boxes walked by on the road. Children cried and dogs barked. I could hear the verses of hallelujah echoing from the evangelical church nearby.

Panti was too excited to sleep. And I was too nervous. He lay in his hammock and I in mine, separated by the curtain. We talked through the doorway for hours before Don Elijio said it was time to sleep.

"You have to sleep if you want to have a dream," he told me. "Remember to pray. Have faith, and the Spirits will speak to you tonight."

His words echoed in my mind, as I curled up around my crystal in my hammock and prayed, whispering my new *ensalmo* over and over and making the sign of the cross over the stone as I had often seen him do.

I wasn't afraid anymore, but I still had a hard time believing that Maya spirits would communicate with me through a tiny oracle that resembled an ordinary marble or a common piece of quartz, pretty as it was. Would the Maya Spirits have anything to say to me?

Did having a sastun mean that I was to become a H'men? I wondered. It was too much to grasp as I slipped off into a gentle sleep.

As the night approached dawn, I felt a powerful tugging on my hammock, jerking me violently from side to side. It was much stronger than I had expected.

Suddenly I was turned upside down, suspended in midair and looking down at the cement floor. I expected to drop out of my hammock at any moment. My heart was racing and pounding as if it too wanted to get up and run away. At the moment of greatest fear and panic, I remembered to say the Our Father, as Panti had told me to do. The hammock swung back down one last time, leaving me right side up, and the dream vision began.

I stood in the doorway of a compact but tidy room. My friend Thomm Noble was sitting at a desk in one corner, motioning for me to come in and sit down. "Come in, Rosita. It's about your crystal."

Across the desk sat a young Caucasian woman with a most pleasant and gentle countenance. She was looking intently into my eyes, offering her quiet support.

I took a seat in front of a small table, which had a glowing kerosene lamp on it.

"Look for a cross of light as you pray into the crystal, Rosita," Thomm told me. "Repeat the Our Father nine times as you make your request. Wash the crystal on Fridays with rum, and keep it near you when you work with sick people. It will strengthen your work."

He told me to look into the flame of the lamp before me on the table. In the light of the lamp, he said, I would see the same shimmering cross of rainbow-colored light that I was to search for in my sastun.

He turned down the flame so that I could look without getting burned. I bent over the lamp and peered into the brightness. I saw the cross take form.

When I regained consciousness in the morning, I jumped out of the hammock. Panti was at his table, waiting for me, and I immediately started telling him the details of my dream.

"The dream was teaching you how to reach the Spirits," he explained as we talked.

For the remainder of the day, we were like two playful children with a secret, delightfully sharing our knowledge of the Spirits who preach in the night.

I also knew what he meant when he promised that when I awoke from a dream vision I would feel like I would live forever. That feeling remains with me, one of the most joyous I have ever known.

CHAPTER SEVENTEEN

Rue   Ruda   Sink In   Ruta graveolens

A cultivated herb found in most gardens of Central America.
It is a panacea, an herb considered helpful in all human ailments.
The aromatic leaves are used for stomach complaints, nervous
disorders, painful periods, delayed or difficult childbirth, and
epilepsy, and rue may be tried for any condition. Healers rely on
rue in the treatment of all spiritual diseases, and it is one of the
three plants that make up the *protecciones*. It should never be
boiled but is rather squeezed fresh into water or tea.

While Panti was in the second hut massaging a patient's belly, I sat at the crate table making amulets. It was stiflingly warm that day, and I'd dragged the table as close to the door as possible in the hopes of catching any breeze, however unlikely, that happened to flutter by.

Now that I had received my sastun, Panti had decided it was time for me to learn more about spiritual illnesses. That day, he put me to work making amulets, which he prescribed frequently to protect his patients against envy and black magic. I had seen him enchant these amulets with his sastun, then tell his patients to keep them close at all times, especially when they left the house or someone they didn't trust came to visit. An amulet's power lasts from about six months to one year before it needs to be reenchanted with the sastun.

Many patients came to him for relief from spiritual illnesses. It hadn't always been that way, but as more people in the region turned to medical doctors for physical problems, his practice had changed. Aspirin and synthetic drugs were available in the remotest villages, so many people didn't bother to seek out traditional medicines for physical problems any longer.

Panti had become famous throughout Central America for his skill in curing spiritual illnesses. Among traditional healers in Central America, there is a hierarchy. There are bonesetters, massage therapists, and snake doctors, who specialize in specific physical ailments. The next level consists of midwives, herbalists, and granny healers like Doña Juana, who are able to treat a variety of physical conditions. But there are very few of the doctor-priests, or H'mens, such as Panti, who in the Maya tradition are able to treat the mind, body, and soul, regardless if the ailment began in the belly or the disquieted soul.

In treating spiritual ailments Panti relied on the same formidable team he used with physical ailments: medicinal plants, prayers, and his sastun. However, the plants and prayers were more potent, and he guarded information about spiritual illnesses much more carefully than that about physical ailments.

The main ingredients he used were holy water from any Catholic church, an herb called Rue, and Copal incense, which he garnered from the resin of the sacred Copal tree.

Rue, known the world around as the Herb of Grace, has many uses in both physical and spiritual healing. It is a favorite household remedy for many ailments, but it is best known for its sure action against evil spirits. The Copal tree is considered a Spirit in its own right and is also capable of canceling out evil.

To this powerful mixture, Don Elijio added a white powder that was made from a calcium-based stone found at a sacred mountain in Guatemala. The stone was called *Piedra de Esquipulas.* It was named for the Christ of Esquipulas, the black Jesus, one of the many versions of Jesus Christ revered in Latin America.

To make an amulet, I placed a sprig of Rue, a chip of Copal incense, and a tiny piece of Esquipulas stone onto a small piece of Balsam bark. I folded a piece of black cotton around the contents, forming a neat little bundle, then sewed up the edges with black thread. It ended up looking like a lumpy pouch, one inch wide and two inches long.

Just as I was tearing a piece of black thread apart with my teeth, I saw a woman run into Panti's yard. She looked agitated and was holding a badly bruised and swollen arm against her breasts. Her arm had begun to turn a grayish shade of blue.

I knew her. Her name was Carla, and she owned a bright blue house we regularly passed on our way to the bush. She always waved at us, holding fresh-cut vegetables from her yard.

"I've come to see the old man because my life is not right," she said breathlessly. Hearing a patient's panicked voice, Panti popped his head out from behind the curtain, then came out to join us.

While she described her symptoms, Panti touched her pulse and nodded his head with increasing melodrama.

A week ago she had been waiting for her husband to come home for dinner, wondering why he was so late. "Suddenly, I felt very frightened, but I didn't know why," she recalled.

As she sat on her front porch brushing her hair and fretting, a warm wind came up from the field behind her house and whipped around to the front.

"The moment the wind blew over me, I felt a chill go through my body. I pushed my children inside and slammed all the doors and windows closed, but it was too late. In the morning my arm was as you see it now. What's wrong with me, *tato?*"

He poked and jabbed at her swollen arm, leaving tiny, white impressions in her taut flesh. She winced in pain and squirmed in her seat. "It is the Hot Wind of the Maya, one of the Nine Malevolent Spirits. I haven't seen that old goat for a long while now," he exclaimed, spreading his leathery lips into an ample grin.

He motioned for me to check the pulse for myself. It was fat and rapid and felt as if it could jump out of its fleshy covering. An icy sensation ran up my arm.

The sastun is not the only way to determine whether an illness has a spiritual or physical cause. Panti also relied on the pulse for diagnosis, ascertaining its intensity and pace. A healthy person's pulse is steady and moderate and is found at the wrist above the thumb. A sick person's pulse can be thin and weak or fat and rapid. The higher up the arm the pulse is found, the more serious the ailment—whether physical or spiritual.

Panti also used the pulse in his treatments. He prayed into it, since he considered it a direct route to the blood, the essence of a person's being.

Carla turned in her chair, pulling at her arm, with tears streaming down her round copper face. Panti began whispering his healing prayers under his breath while blowing tenderly on her arm.

"Fill her bag with *Zorillo,* Skunk Root, Rosita," he ordered, then turned his attention back to Carla. "Start drinking the *Zorillo* tea tonight while saying an Our Father. Rosita and I will bring the *Tzibche* plant to your house tomorrow. We'll add it to the Xiv we're mixing up now. You'll use it for nine herbal steam baths."

Reaching under his cluttered table, he pulled out a hand-sewn cloth bag, which I recognized as Chinda's handiwork. From the bag, he removed a handful of dried, powdered Copal resin. He placed it on a piece of paper that once belonged to a child's exercise book and was covered with the repetitive strokes of alphabet letters. He rummaged around in another cloth bag and removed a slice of stark white, sticky Copal resin from a ball the size of a man's fist. He added the resin to the powder, then shook in dried Rosemary leaves.

I smiled. Panti was always sifting through gunnysacks. Sometimes he looked like a tropical version of Santa Claus, with his bag of roots, vines, barks, and leaves. They were his treasured presents from the Maya Spirits and he was jolly old St. Nick, right down to his boisterous chuckling.

He wrapped up the mixture and tied it up with a single strand of plastic he had ripped off an old flour sack. "Burn this incense on coals in your house every Thursday for nine weeks and outside your house on nine Fridays. Say this prayer: All evil and envy should leave this place now because it is causing much harm. In the name of the Father, the Son, and the Holy Ghost. Amen."

With this part of his work completed, Panti proceeded with the comic entertainment. Several times he made her giggle, despite the traces of tears on her cheeks.

Eventually she clutched her bag of herbs and started toward the door, muttering to herself, "Praise God. Praise God. God bless you, old man, with a long life to help us who have nowhere else to go. There's no one like God and Don Elijio."

He shrugged modestly and smiled at her. "Have no worry, child. Soon you will be well again. God will help us all."

After Carla left, Panti turned to look at me with his eyebrows bent across his brow in warning. "You see, *mamasita,* that is why the people in these parts always sleep with their doors and windows closed tightly. The Winds favor no one."

The comment was intended to convince me once and for all not to complain about him closing up the windows and doors at night so not a whiff of fresh air seeped in. With the tropical heat lingering into the night, though, the cement house was often unbearable for sleeping. But he especially feared the Night Wind. "It is pure Spirit and can harm the people it touches."

Although he'd told me before about the Winds, Carla was the first case of physical symptoms that I'd heard blamed on them. The Winds are intelligent entities to many Latin American cultures, known as *Ik* in Mayan and *Mal Vientos* in Spanish. "They are especially feared because they could make people physically ill with mysterious ailments for long periods of time," he said.

"Some people don't believe in them anymore. Even when they're sick with them, they still don't believe," Panti continued. "I respect the Winds because I know how powerful they can be. I've seen their mischief."

He said one of the worst is the rain-drenched Wind of the Milpa, which attacks farmers on the way home from the cornfields. "They're hot, sweaty, and tired, and the cold, wet wind blows on them. The next day they can't get out of bed, won't eat, and have a high fever."

Some Winds do not intend to do harm, but people must show respectful caution anyway, he warned. "Avoid them whenever possible. Farmers should carry dry clothes and get under a shelter when it rains in the *milpa*."

We continued to talk over lunch. I asked him why the *Tzibche* plant was so vital to Carla's mixture. He had used it during my initiation *Primicia* to protect us from being harmed by the powerful, spiritual forces that were roused.

"It is the only Xiv that cures the Hot Wind of the Maya. It grows just behind a tree near the peanut field on the way to the forest. Tomorrow I will show you."

I then asked if there were any other spiritual diseases I should learn about. He looked at me as if I'd asked him how many shades of green there were in the forest.

"The human spirit can be plagued by as many troubles as the body," he said. Infants were particularly susceptible. A child was not just cranky when it refused to eat, was up all night, and cried often. He outlined three illnesses the child might have: *mal de ojo, susto,* and *viento de descuido.* Their symptoms are similar and they are distinguished only by the intensity and rapidity of the pulse.

I knew of *mal de ojo,* or the evil eye. It had been brought to the New World by the Spaniards, who probably picked it up from the Arabs. In my childhood household, my Italian grandmother, Isola, had insisted I wear a clove of garlic in my Catholic scapula to protect me from *mal occhio,* the evil eye in Italian.

Susto was Spanish for fright. It can be caused by anything that might scare a baby: an angry dog, the piercing sound of a jet engine overhead, or a drunken father's abuse of its mother.

Viento de descuido means Wind of Carelessness. After questioning mothers, Panti traced infants' problems to being left near a drafty window in the early morning hours or taken outdoors at dusk with their heads or bodies uncovered.

As usual, the afternoon transport arrived just as we finished lunch and another discussion of Carla's *mal viento.*

An attractive young woman in a miniskirt walked in, followed by a square-framed, middle-aged man.

The young woman had traveled all the way from Guatemala City to see Panti. Her brother had been abducted by the Guatemalan military a year ago. Her family had never heard from him again. He had become one of Guatemala's many *Desaparecidos.*

She wanted to know if her brother was still alive.

"This is the work of the sastun," Panti said as he guided her into the cement house.

He sat at the table and she sat on the stool close to the door. He pulled the little clay jar containing the sastun out of the Ovaltine can and asked the boy's name.

"Ricardo," she said.

He turned the jar upside down into his left palm, and the sastun fell out. He blew three times on the sastun and three times into the clay jar, then placed the sastun back into the jar.

He twirled the jar with the sastun inside in circles. It made the clacking sound I now knew well. As he twirled the jar, he sang a Mayan chant.

He dumped the sastun out of the jar into her right hand and instructed her to hold it like a die and shake it.

After a few minutes, he directed her to the doorway where the light was better. He opened her palm, poked at the sastun, and peered into it, searching for the answer.

He motioned for me to come over and showed me with his finger a number of tiny bubbles inside the translucent ball.

"There it is, there it is. Do you see it?" he said. I saw the bubbles, but it was like a foreign language to me. Panti could see a meaning within those bubbles that I couldn't.

"The boy's luck is good," he said. "He is alive but far from home."

"Can you bring him home?" asked his sister.

"Did you bring a photograph?" he asked.

Then she pulled out a tiny black-and-white dog-eared photograph. I peeked at the picture and saw a very young man with sweet eyes staring up at me.

Panti placed the photo face down on the stained plastic tablecloth.

He twirled the sastun in circles around the photograph, repeating a Mayan chant.

"Sastun, sastun, with your great power," he sang and went on to ask for the boy's safe return.

Panti gave her back the photograph and instructed her to place it upside down in a pocket over her heart every Thursday and Friday and repeat, "You are mine, come here, sit down, and stay."

The young woman went to sit outside and wait for a ride as I motioned for the square-built man to come into the hut.

He had greasy black hair and a sallow complexion. He looked as if he hadn't shaved or bathed in a few days. His clothes were rumpled, and a fungus-ridden toenail escaped from a hole in his sneaker.

"What is your problem?" asked Don Elijio.

He hesitated, glancing at me.

Before he had a chance to say a word, Don Elijio said forcefully, "She is with me. What I say, she says. What she says, I say."

The man shrugged and pulled out a small photograph of a pretty young girl from his wallet. She looked young enough to be his daughter.

"I want this girl for my own," he muttered. "I had her father's permission to see her but then she changed her mind, just before our wedding. I want her back. Can you help me?"

Don Elijio picked up the picture and turned it over. "Some people are lucky with women," he said. "I've been alone for many years and will probably die that way."

He enchanted the photo with his sastun, handed it back to the man, and instructed him to place the photo upside down in his pocket every Friday for nine weeks, repeating, "You are mine, come here, sit down, and stay."

The man paid him five dollars and left hastily.

I watched him disappear down the road. As soon as he was out of earshot, I turned to Don Elijio and asked, "What was that all about? Do you enchant women often for men?"

"Yes, *mamasita,* I do it all the time," he said matter-of-factly. "The *encanto* lasts for six months only. During that time he must prove himself worthy of her. She comes out of the spell with an angry fury and will only stay if he has been good to her. It only gives the man a chance."

"By then she could be pregnant and stuck with someone she doesn't love and never intended to marry," I said. "I don't like it. Do you enchant men for women?"

"If they ask," he said. "But few ever do. Women are more sensible than men, you know."

I let the subject drop. I knew that H'mens had performed romantic enchantments since ancient times. I was sure that I never would; it was against everything I believed as a woman. Had this happened earlier I might have been scared away, but I knew by now that Don Elijio had a pure heart, and a lot to teach me. I was here to learn. I returned to my task of making cough syrup out of fresh mango, avocado, guanabana, and cotton leaves.

After a light dinner of beans and tortillas, Don Elijio announced it was 7:30 and time for bed. Without another word, he pulled the windows tightly shut and sealed himself behind his battered wooden door.

As the purples of dusk suffused through the coral sky, I thought about the question of faith. A burn is soothed by an Aloe plant, regardless if the recipient believes in its medicinal value. But with spiritual illnesses, faith in the healer and the cure being offered had a lot to do with the patient's recovery.

Still, some of Panti's patients complained of feelings and symptoms that were foreign to me. I didn't know what I would say or do if a patient blamed the wind or told me that an enemy was taking revenge through black magic. It was a big leap of faith for me to accept that evil wishes could cause physical or spiritual harm. Or this pleasant evening breeze.

I lifted my face toward the sky and smiled, feeling the cool wind against my grateful body.

CHAPTER EIGHTEEN

Skunk Root   Zorillo   Payche   Chiococca alba

A tropical vine whose dark, brittle, and meandering root is
one of the most important remedies in Maya medicine. The root
is used for stomach ulcers, pain, and ailments with multiple,
confusing systems. The bark of the vine and the root are taken as
a tea to ward off black magic, envy, the evil eye, and evil spirits.
The leaves are part of the Nine Xiv formula for herbal bathing.

Only a few weeks later, Don Elijio and I were making a wound powder
of dried Cancer Herb and *Tres Puntas* leaves when a stocky Indian man
bolted through the doorway.

"Don Elijio." The man's voice trembled. "I come with much faith. I
bring you a woman who is truly ill."

The man stepped to one side, revealing a disheveled woman in her
late forties wearing torn and badly soiled clothes. Her hair was stiff and
matted, her complexion gray, and her expression tortured. It was her
eyes, though, that were the most frightening. Her fixed gaze sent an icy
feeling through my veins.

She was being propped up by a brawny man on either side of her.
Her strong escorts, we learned, were her sons Roberto and José.

I instinctively grabbed her hand and helped her to sit down on one
of the stools. She slipped forward, drooling uncontrollably. The drool
streamed down her chin and fell onto her lap. The son named Roberto
quickly jumped up to wipe off his mother's face.

The woman began shifting her body from side to side and doing
what I can only describe as a panting growl. I looked over at Panti to see
he was as calm and in control as ever.

"Our mother took sick about two weeks ago," explained José.

Late one Tuesday night, their mother, Angelina, was in her room combing her hair while the rest of the family sat in the living room. Suddenly they heard her door slam shut and the lock click. A coarse, unfamiliar voice bellowed from her room, then moaning and a blood-curdling cry, followed by the sounds of shattering glass and chairs being thrashed.

They were horrified. They screamed for her to open the door, but she didn't respond. When her husband finally broke the door down with an ax, the family rushed in to see that nearly all her belongings had been broken and strewn about on the floor. Angelina was bleeding, tearing up the sheets, and biting herself on the arms. It took six adult men to hold her down on the bed and restrain her.

While her son relayed the gruesome details, she continued to drool and rock back and forth on the stool. From time to time she opened her mouth to try to speak, but the sounds that came out resembled the cries of a wounded animal.

At one point, I thought I heard her mumble, "Help me, please," as she reached out for me, clutching my sleeve and twisting it with all that was left of her ebbing strength.

I wanted to run, but I took a deep breath and let her hold me. A disturbing cold chill traveled down my spine. The hair on my arms stood straight up as if electrified.

I looked at Panti again. He was listening intently, tapping his lips with the four fingers of his right hand as was his habit when listening to patients.

"A *curandera* came to see her the next morning. She said prayers for her and gave her medicine to drink that calmed her down a bit. But within the hour she was terrible again. The *curandera* stayed with her all night long praying and giving her teas, but after five days she said she couldn't help anymore. She said my mother was possessed by an evil spirit that was too powerful for her prayers and medicine."

The *curandera* told them to take her to Don Elijio, who could not only subdue the bad spirit but chase it out of her body.

"Don Elijio, my mother has not closed her eyes to sleep for fifteen days now, nor has any food passed her lips. We fear she will die in this awful state. We have faith in you. Please help her, *señor,*" said José, wiping tears from his eyes and resting both his hands on his mother's shoulders.

Panti stood up, and all eyes but Angelina's followed his every move. He motioned for me to go outside with him.

"Put some water on to boil immediately. Put *Zorillo* in there and get it boiling good for ten minutes and bring me some hot coals in this can as soon as possible," he said swiftly, turning on his heels to get back to Angelina's side.

I scooped out a handful of *Zorillo* into a gourd bowl and carried it to the kitchen. As I plopped the foul-smelling root into a pot and started a fire under it, my hands were trembling.

I rushed back to the cement house with the coals, catching Panti dousing the woman with holy water from head to foot, nearly shouting the Mayan prayers over her.

I went to fetch the pieces of Copal incense Panti had requested. He tossed them onto the coals in the tin he'd situated under her stool, and the rich incense enveloped her. He also sprinkled dried Rosemary on the coals as he repeated the prayers and held her pulse.

He joined me in the kitchen to check on the boiling *Zorillo*. I put another piece of wood on the fire. I looked over at him for reassurance. I felt frightened by the woman.

"I am sure it is black magic, Rosita, but I will ask the sastun to make sure." Shaking his head, he added, "It is awful what people do to their neighbors."

To acknowledge that this poor soul suffering right in front of me was a victim of black magic was too much for me to accept. So I concentrated on her physical symptoms. Whatever troubled her, she was obviously in need of Don Elijio's herbs and prayers.

I cooled the *Zorillo* concoction by pouring the dark, pungent liquid back and forth from one gourd bowl to another. I could hear Panti twirling his sastun, repeating her name along with words like *espiritu maligno* (evil spirit), *maldad* (evil), and *hechismo* (black magic).

I carried the cooled tea into the house, and Panti started mixing in Rue, holy water, and the sacred Esquipulas stone. He handed her the mixture in a gourd, and I half expected her to bat it to the ground. Instead, she steadied it to her lips and drank it down, sip by sip. It was the only cognizant act I had yet seen her perform.

She soon made a motion that she had to vomit, and Panti and I lifted her by the arms and led her outside, where she held onto the trunk of the Sour Orange tree. She wretched and moaned and uttered incoherent words, as if she were mumbling in a foreign language. As she threw up phlegm and the foulest fluids I'd ever smelled, I could feel my face get hot, while my heart beat loudly in my ears.

Panti spoke quietly to her as she gagged and spit at the muddy ground. "That is very good, *mamasita*. You will be well soon, I promise you." He marched back to the house, yelling over his shoulder, "Stay with her, Rosita."

The repulsive odor was overpowering my senses and I wanted to bolt again. Yet something held me at my post. It didn't matter if I believed in black magic or not. She was a sick woman who needed my help.

I held Angelina's shoulders and patted her head as Panti asked her sons about her enemies. Any jealous neighbors or angry relatives? he asked. They mentioned their sister's former boyfriend. Angelina had recently rejected him as a suitable husband for her daughter. The boyfriend's mother had paid Angelina a visit on the Tuesday before she took sick. The woman had angrily accused Angelina of assuming her daughter was too good for her son. As she stepped off their doorstep to leave, she had cursed them, shouting, "Soon you will see suffering come to this house because of your pride."

Don Elijio continued his probe. Were they missing any personal items or photographs? They are often used in spells. Did they find any *pachingos*, voodoo dolls, buried in the house or hidden inside rooms? Panti instructed them to look through their mother's belongings and dig around the doorstep for any *pachingos*. Then Panti's sastun confirmed that the boyfriend's angry mother had paid an *obeha* (black magic) person to cast an evil spell over Angelina.

With that news, the sons exploded, waving their arms about and pleading with Panti for revenge. "We want to see that disgraceful woman suffer herself for what she has done to our mother. We will pay you whatever you ask."

Panti put his hands into the air with the palms up as if to stop an oncoming blow, warning, "I will cure your mother. That I can do because God has given me these powers. But you must go to another house to bring harm to someone."

I helped Angelina limp back to her seat, wiping her face with a cloth I'd wrung out in cool water. A tear fell off her cheek and onto my hand, and with that lone tear, all my fright evaporated, absorbed by that single drop of emotion. I suddenly felt ashamed that I had ever been scared enough to think of running.

Panti continued to soothe her, telling her sons that she could go home after he treated her for another hour or so. He seemed satisfied by her improving condition, enough to relax a bit and sit down in his customary chair.

How did he know it was black magic? I asked. "The symptoms are in the story," he explained. "The boy's mother announced her intentions when she stepped off the doorstep with that curse. And evil things usually begin on Tuesday." Also, that her pulse was fat, jumpy, rapid, and high up on the arm was clear evidence.

"There is no lack of people to do these evil things, but there is a lack of people to cure them," he said to everyone as we gave the woman, much subdued, another draft of *Zorillo,* which she eagerly downed in just a few gulps. He showed the sons how to mix up the concoction and also instructed them how to place the Copal mixture under her seat so that she would get the full effects of the billowing, healing smoke.

"Close all your doors and windows this Thursday and carry the smoking incense around the rooms, saying nine times, 'All evil leave this place as it is doing much harm. In the name of God the Father, God the Son, and the Holy Ghost.'"

Panti said it would take nine days to heal her. "It will take that long to fully cleanse this malignant spirit and its effects from her body, " he said. "It is a very bad one and may be the devil himself. Sometimes he likes to do the work himself, especially if the victim has led a very good and religious life. That person is a special prize for the Prince of Darkness. But have no worry, my sons, God is on my side."

"Don Elijio, are you not in danger in this work? How do you protect yourself from being killed or harmed by these bad spirits?" asked Roberto.

"I was *curado* by my *maestro.* No one can harm me because I am protected by his spirit." Jerónimo had said the sacred special prayers for him and had given him protective teas to drink for nine Fridays.

"No, try as they may, they cannot touch me. They get very angry, I know, because I undo their evil nets of filth and greed every day. And, sadly, there is only me left to fight against many who know how to do this evil."

He believed the use of *maldad* had increased greatly over the years, especially after a how-to book was brought back from the United States. He had first seen the book in the 1950s, with its pages full of evil spells and incantations.

"Right there it tells you how to do all manner of evil against your neighbors. Prayers to the devil using frogs and other animals. I can't catch my breath, I'm working so hard to dig up the *Zorillo*. Thirty years ago, I might have used a few pounds in a year. Now, I need a hundred pounds a month. How long can it last if I keep digging up so many roots every year? What will we do if the *Zorillo* runs out?

"Men have become selfish and greedy," he scolded. "They have forgotten the Maya Spirits and stopped doing their *Primicias,* which used to protect them."

Angelina continued to vomit from time to time, and the dark, steamy cement house, cooking under the tropical sun, reeked. All Panti said was, "Good. You need to cleanse yourself of this filthy being. He is inside you right now and frightened, looking for a way to escape. Soon it will leave and you will be yourself again," he said.

"How much do we owe you for our mother's life, *papá?*" asked Roberto with tears in his eyes.

Panti sat down in his chair, shiny from years of use, and whispered, "Whatever you would like to give me." They handed him the equivalent of one hundred dollars.

Ten days later, Roberto and José brought their mother back to Panti for a checkup. I couldn't believe the transformation. She was clean and neat and well dressed. She smiled sweetly and seemed entirely normal.

Her sons stood in the doorway, washed in sunlight, and humbly said to Panti, "*Tatito,* we have returned with our mother to kiss your hand and to give thanks to you and to God."

Angelina stood just a pace behind them, then came forward to reach for Panti's hand, which she kissed according to the old Latin custom.

"Thank you, *señor,* for my life," she said graciously.

"It is just as I said, these evil spirits have no real power," Panti affirmed as he felt her wrist, whispering nine prayers into her pulse for her continued well-being.

Her sons said that once they got her home and she began to recover, she couldn't recall her behavior.

"But still she slipped away from us at times, drooling and talking incoherently. Then my sister spent the night with her one Friday to pray. She burned Copal all night, and for the first time since she took ill, we didn't have to tie our mother down on the bed."

After the seventh day Angelina had fully recovered but her family had continued the treatments for two more days as Panti had ordered.

Panti told me to feel her pulse. I felt a slow, steady beat characteristic of a person in physical balance and health.

It was fortunate, I thought, that her sons had brought her here. If she'd been delivered to a more conventional clinic, she probably would have been drugged and restrained, languishing the rest of her life in a mental institution where no one would have considered the possibility of demonic possession.

I asked him later how he discerned the difference between possession and madness, which he acknowledged was rooted in natural causes and psychological dementia. "Had she been mad," he answered, "I probably wouldn't have been able to help her as easily. Madness takes much more time to heal and sometimes is incurable."

As Angelina and her sons left, Panti remarked that madness was caused by thinking too much. People go mad fretting over circumstances for which they have no control—an unhappy past or a doomsday future. No psychology book could say it much better than that.

CHAPTER NINETEEN

Duck Flower   Contribo   Aristolochia trilobata

A major medicine of traditional healers throughout Central
America. Taken as a boiled tea or soaked in water it has a
wonderful effect on gastritis, fever, colds, flu, constipation,
and sinus congestion and is often drunk in rum to alleviate
hangovers. Now disappearing, it is a medicinal plant
much in need of protection.

My first warning of trouble was on a Saturday morning shopping trip to
San Ignacio Town.

Crystal and I were in the Venus Store buying school supplies, when
another shopper casually remarked, "Oh, I was going to see Don Elijio
about my crippled daughter, but the evangelist healers are coming tomorrow so I'll bring her to them to be cured instead."

I didn't think much about it. Many Central Americans had converted to Protestant Evangelism in the great waves of revivals that swept
through the region in the 1950s. Every so often a group came to Belize
for a few weeks, fanning across the country, sending out preachers to
towns and villages with promises to heal body and spirit. They brought
with them generator-operated loudspeakers to broadcast their meetings.

But when I arrived in San Antonio the next Wednesday for my regular three-day stay with Don Elijio, it was clear something had changed.
Hardly anyone was out on the road or on their doorsteps chatting with
neighbors. A familiar face by now, I was usually greeted by a chorus of
children, smiling women, and barking dogs. Today, only the dogs ran to
meet me.

I found Don Elijio sitting alone inside his cement hut. It was un-
usual for him to be alone this time of the morning. If he wasn't out in

the forest, I'd find him visiting with an old friend, neighbor, local patient, or a patient who had moved in for treatments. On rare occasions I had found him chopping alone, or fidgeting and lonely. But today was different. I could tell by his face that something was amiss.

"No one has been here to consult with me all week," he spilled out with anguish as I put down my bag.

He hadn't had any patients yesterday or the day before. On Monday only a Guatemalan woman had come. Tuesday, a man from Belmopan had brought his feverish son. This was in contrast to his regular weekly schedule of around a hundred patients.

By the time I met Don Elijio in 1983, he had built up a thriving practice all based on word of mouth. Some days he had thirty patients, but there were days when only a few patients came and some when no one arrived at all. Those were sad and lonely days for Don Elijio, and boring for me. He'd sit at his crate consumed with uncontrollable thoughts of rejection. At every sound, he'd start, hoping a patient would appear at the door. In the hopes of distracting him, I'd read aloud herbology books in Spanish.

"My patients have abandoned me," he cried. "It's those cultists. They've thrown me over in a flash for those cultists."

He was talking about the evangelists. They had arrived on Sunday and were conducting a week-long revival meeting down the road in a community building in the village of Cristo Rey. As their evangelism was revived, the people discarded their Catholic customs, fiestas, saints, and beliefs in exchange for a simpler doctrine: Jesus is the only way, worldly ways lead to the devil.

It wasn't just the Catholic saints and the Virgin of Guadalupe who were discarded. Ix Chel and the Maya Spirits were doubly condemned: Evangelism required a complete rejection of the old Maya ways and beliefs.

So when the evangelists conducted a week of soul saving for eternal life with Jesus in paradise, Don Elijio was a very lonely man. He was, sadly, an anachronism twice over since his healing was based on his friendly combination of Maya and Catholic lore. His old brand of Catholicism was accepting and had co-opted many Maya beliefs in order to survive in the people's hearts. And Maya religion had also co-opted Catholic beliefs in order to survive. There was no intolerance in Don Elijio's heart.

"Every night I can hear the loudspeaker going in Cristo Rey—the shouting and the screaming," he wailed. "No, they don't even want to hear my name now. Only hallelujah, hallelujah, brother, and pass the donation basket.

"Those healings that they do don't last," he added. "I've seen it with my own eyes. They flail their arms about, they faint, they holler and get up walking or cured, but in a few days after the screaming and shouting is over, the sickness returns. Then, yes, they want to look for Don Elijio again.

"They don't understand," he said. "It is their own faith that heals them, not the evangelist preacher."

It didn't take much to become an evangelist preacher. This was part of the appeal of the movement. After a few months, any man or woman could become a preacher. By contrast, it required years of training, as well as a vow of celibacy, to be a Catholic priest. And as I was learning firsthand, it wasn't any easier to study to become a H'men.

The evangelical movement was one of the major reasons why so few of the younger generation learned the old Maya ways. Many members of Don Elijio's family had converted to Evangelism over the years, and those who had could barely tolerate their family patriarch.

He was inconsolable. "They've forgotten me," he wailed again and again. I gave him the gift that I had brought with me—a bottle of wintergreen oil. I told him I loved him and would always be at his side. I gave him a treatment, rubbing the oil into his sore muscles.

After the treatment he cheered up a bit, but I was still concerned. He needed his patients—as much, surely, as they needed him. His patients were his family, his companions, his audience, and his reason for being. Without them, he was devoid of purpose and direction and mourned for Chinda more than usual. The energy they gave him explained why he could treat as many as thirty patients in a day and at the end of the arduous administrations feel better than before.

It was the same for me. In my healing work, I too noticed that if ever I began a treatment or consultation feeling tired or drained, I was always renewed and strengthened afterward. I knew this was God's gift to the healer—that your patients strengthen and heal you as well as the reverse.

We sat alone chopping plants in the late morning. To cheer him up I told him that I'd been approached by the producers of Belize All Over, a new local television program. They wanted to make a documentary about him and his work, which would be the first in a series about life in Belize.

Now, at ninety-two, Don Elijio had never seen television and had no idea what a documentary was. I explained as best I could. Television had only come to Belize in the late 1970s, and since San Antonio still didn't have electricity, television was not a part of daily life. I told him I thought the documentary was important so that future generations of Belizeans would know him, what he looked like, and how one man who never went to school had become more sought after than a government minister.

"Yes, yes, that is true," he nodded. "Bring them when you like," he said and went back to his gloom and chopping.

After lunch, the afternoon transport came in, Angel at the wheel, with Isabel and four of the younger children all squeezed into the front cab.

I was thankful when I saw a familiar face appear at the door. It was Doña Rosa, laden with her bags of wares.

Doña Rosa had been one of Don Elijio's first patients when he was just starting out, back in the days when Chinda was still alive. She was a gregarious woman, short of stature, square framed, and full of laughter and stories much like Don Elijio. She was a trader, as Don Elijio had once been, and they had developed a deep friendship over the years. Whenever Doña Rosa, who lived in Benque—a town near the Guatemalan border—came to town to San Antonio to sell dresses, pots, pans, towels, and cosmetics, she stayed with her old friend, Don Elijio.

The two of them were wonderful company for each other. Doña Rosa was four decades younger than Don Elijio, but she was part of the old world too. They loved to tell each other stories of the old days, and I had come to love her as much as he did. She doted on him. She cooked his favorite dishes on the open hearth and brought him vitamins, eye drops, and dried herbs from Melchor, the border town on the Guatemalan side.

Doña Rosa lived in Benque with her husband Poncho and two teenaged daughters. When on rare occasion Don Elijio left San Antonio, he could be found at Rosa and Poncho's home in Benque, where word would spread of his presence and patients would line up outside the door holding sick babies and propping up infirm grandparents to receive his prayers and herbs.

Recently, Rosa had begun to learn about the plants and prayers so that she could do a little bit of healing in Benque. She learned fast, but she didn't like to deal with evil spirits, nor would she take on chronic or severely ill patients. She wanted to become a granny healer and refer all serious cases to Don Elijio.

Don Elijio's face lit up the moment he saw Doña Rosa at the door. He immediately went into the routine he always said when both Rosa and I were in San Antonio at the same time.

"Ahh, I have a rose garden today," he said.

We joked about him being a thorn between two roses, and he shamelessly responded, "Get closer, both of you, and see if you can make me bleed. I have lots of blood."

Doña Rosa spent the night and we had a cheery time, despite the distorted drone of preaching, blaring through scratchy speakers five miles away.

But in the morning, I could see Don Elijio was suffering again. He didn't want to go out in the rainforest, he was well stocked. What he needed, he said, was patients. To add insult to injury, some village men, fresh from the revival meeting, came by and asked Don Elijio to give up his practice.

"This is the devil's work," they told him.

That got his attention. Don Elijio bolted upright in his seat and said, with great force and indignation, "You are wrong. I have no pact with the devil. I work only with God and the Nine Maya Spirits. Devil's work is evil. My work is healing. Never has anyone walked in here and had to be carried out. But many were those who were carried in and walked out."

The three men shifted uncomfortably in their seats. I recognized at least two of them as former patients.

"But I tell you what, let's make a deal," continued Don Elijio, sounding strong and in control.

"Just give me twenty-five dollars a day of your daily collections to live on. Some days I make up to two hundred dollars. But I'm not greedy. I wouldn't ask you for that much. Don't ask me to sing or read the Bible, because I can't read. Don't ask me to clap my hands and stamp my feet, because I have rheumatism. If you agree to this, then yes, I will give up my work."

They left indignant and never bothered him again.

Fortunately, right after they left, Doña María, the wife of Manuel Tzib, stopped by. At fifty-seven, she had delicate features with sparkling eyes and long eyelashes. A slim woman with a youthfully thick gray braid down her back, she wore a homemade, faded, cotton dress under a brightly colored apron.

Her husband, Manuel Tzib, was Chinda's uncle and one of the original settlers of San Antonio at the turn of the century. He had contributed his knowledge to Don Elijio when the younger man decided to become a healer.

Tzib was still alive at 107 and lived just a short walk from the clinic. He spent his days bossing Doña María around from his hammock, wrapped in his wife's *reboso* (shawl) and wearing her plastic shoes on his tiny feet.

Doña María had married the widowed Tzib when he was sixty-five and she was fifteen, over the objections of both families. But now, decades later, she was the family member who watched over Don Elijio most closely. Before Chinda died, the two women had made a pact: they would care for each other's husbands in the event one of them died.

So it was Doña María who lovingly laundered the clothing Chinda had made Don Elijio before she had died. She stopped by twice a week to pick up and drop off his laundry, bringing along treats of tamales and sweet buns. In return, Don Elijio helped support her and Manuel.

After she folded the laundry and swept the floor, she sat down to chat. Her husband, who never left the house, had a cold. Don Elijio immediately set about filling her apron with Contribo, Duck Flower, vine, instructing her how to prepare a special sun tea for the elderly.

"Everybody has a cold right now," said Doña Rosa. "My husband and my daughters are at home right now, sniffling and sneezing."

"These are natural things," said Don Elijio.

I concurred. "God willing, we never find a cure for the cold," I said, "because it is a way the body has of cleansing itself regularly. If we take it away, more serious diseases will follow."

"It's good for all that gunk to come out of the body," threw in Don Elijio.

Doña María left for home, making a joke about her jealous husband in his hammock. A few days later, Doña Rosa and I left together, worried about how Don Elijio would fare alone. Not that he was entirely alone. Angel, Isabel, and their eleven children were always around. Of all his grandchildren, Angel was the one who took the most responsibility for his grandfather.

Still, it was a tough few weeks for Don Elijio. The making of the documentary was well timed. The film crew and I met up in San Ignacio and headed out for San Antonio, where we found the old H'men in rare form, ready to be a star.

I remember how he asked me to comb his hair and help him dress in something appropriate. He was relaxed, charming, and powerful on camera, as if he had done it a thousand times before.

When the revivalists pulled up stakes, his patients—as he had predicted—came back one by one, sheepish looks on their faces. Even one of the men who had asked him to give up his practice showed up one day with a painful, swollen jaw.

When the documentary aired the following month on Belize television, it was a huge success. Because of its popularity, it was played over and over again.

Once he was on television, his fame spread into parts of the country where he had been little known. For a while, twice as many patients showed up at his door. Many of the new patients spoke little Spanish, and it became necessary for me to spend more time in San Antonio as his assistant, translating, chopping, collecting. Sometimes Greg came to help out, lending his skills and expertise. Even with the three of us, there were times when we could barely handle the influx. Doña Rosa filled in when we were absent.

In spite of my efforts to bring Don Elijio to town so that he could see the documentary in the home of the only person I knew who had a television, it was a full year before he had a chance to see it. Doña Rosa bought a TV, and that same weekend Don Elijio went to visit.

Like excited children, they turned on the TV for the first time, only to see Don Elijio sitting at his chopping block, discussing Man Vine and *ciro*.

"*Ciro* is something that jumps in your belly like a rabbit," said the Don Elijio on the television.

When he came home, he told me how impressed he had been. "It remembered everything we said that day," he said proudly.

CHAPTER TWENTY

Wild Coffee   Café Sylvestre   Eremuil
Malmea depressa

The aromatic leaves of the small tree are considered to be
the most important of all the medicinal leaves used in Maya
medicine. Traditional healers use the leaves, boiled in water, to
bathe patients suffering from any sickness. A steam bath of the
leaves is used to treat muscle spasms, rheumatism, arthritis,
paralysis, swelling, backache, and fever.

It was a delightfully cool, sunny winter afternoon in late December
when a truck pulled into the driveway of Ix Chel Farm. Rolando, our
employee, came to get me in the garden where I was laboring over a
bed of collard transplants and daydreaming about the salads and pots of
delicious boiled greens I hoped we would be eating in a few months.

"There's a man here to see you, Doña Rosita," Rolando said softly,
so as not to startle me out of my reverie.

"Did he say what he wants?" I asked, reluctant to leave my garden,
thinking I might never finish the transplanting.

"He said he has brought a sick child."

"Ask him to wait on the veranda, get him a drink of water, and tell
him I'll be right there as soon as I wash my hands," I said.

A couple in their midthirties sat in chairs on the open-air, thatch
roof porch that Greg had just built. The woman held a girl, about eight
years old, in her arms. She seemed stiff as a board sitting in her
mother's lap. Her outstretched legs were oddly askew. I braced myself
for the story.

"Good afternoon," I said. "I am Rosita. What can I do for you?"

As the man introduced himself, I recognized him as the manager of the hardware store in Santa Elena, the town just across the Hawkesworth Bridge from San Ignacio.

"We've come to see you because our daughter is very ill," he said, "gravely ill, and no doctor has been able to help her. She hasn't been able to move her legs or walk for more than three weeks now. The doctors sent us home saying they didn't know what the problem was and had no treatments to offer. As we left, one suggested we find someone to massage her. If I were a rich man, I would take her to Merida, Miami, or Guatemala City, but I am poor and anyway, I've heard of you. People say you do good work and that you learned a lot from Don Elijio. I come to see you with great faith, Doña Rosita."

"Tell me what happened," I asked.

The mother, an attractive woman, who like her husband was of Spanish descent, took a deep, troubled breath. She let out a mournful sigh.

"About two months ago, Shajira [Sha-heér-a] had the flu," she said. "It was a very bad case and she was home from school for what seemed too long of a time. Usually my children recover quickly and well from colds and flu, but this time Shajira stayed in bed for three weeks. Then one morning, she was unable to get out of bed. She said her whole body hurt terribly and she was unable to move from the waist down.

"We took her to the hospital in Belmopan," she continued. "They sent us to Belize City, where they kept her for observation for ten days. They said that none of their tests could tell them why she was unable to walk. The doctor prescribed some pills, which made her vomit, so we stopped using them. Then he gave us some pills for the pain, which made her feel nauseated and sleepy. We didn't like that but were afraid to stop the medication and leave our daughter in pain. So we started giving her aspirin, but that only helps a little.

"Now, it has been three weeks since we came home from the hospital with her, and we see no improvement. She cannot walk and there is little feeling in her body from the waist down. Some days she has a fever, no appetite, and seems listless. And she cries often saying she's afraid to be a cripple. Can you help us?" the mother pleaded.

I took a long deep breath and looked searchingly at little Shajira. She was exquisitely beautiful, with a warm terra cotta complexion, shiny black hair, big round doe eyes, and a sweet expression.

"Bring her into the examining room and let's have a good look," I said. Unfortunately Greg was back in the States visiting his parents, and I missed his comforting presence. We usually teamed up for difficult or seriously ill patients.

I examined the child and determined that there was soreness but no swelling in the tissue of her upper back, very close to the spinal column. Her lower back and the entire musculature of her spinal column were tense and rigid.

Shajira responded nicely to tickles and pinches, which was very encouraging. But every time I put pressure on her right upper back she screamed out in pain and tried to wriggle off the table. Moment by moment I was feeling more confident that I would be able to help her. I wasn't sure how much of the function of her legs and torso she would recover, but I felt the general prognosis was good.

It had taken me many years to be comfortable enough with Don Elijio's healing techniques and plants to incorporate them into my own practice. But here I saw a perfect opportunity to blend his system and mine.

I told Shajira's parents that I thought a pocket of infection left over from the flu virus had settled in her spinal column at the very point of a nerve plexus. This, I thought, was causing her paralysis.

"She will need two naprapathic treatments each week for a while and a series of steam baths in between the treatments," I counseled. "I would also like her to take something for that virus and have a good purge.

"Are you willing to follow this therapy?" I asked the worried parents.

"Doña Rosita, we have no other hope," said the father. "Thanks be to God that you are here and there is hope for our daughter. We have four other children who are all strong and healthy. Shajira too has always been a healthy child. I can't bear to see her like this. She sits in the window looking down on her brothers and playmates, longing to run and play. Yes, we agree. Of course we do. We will do whatever you say."

"Good, then," I answered, pleased to have their cooperation and understanding of the task before us.

I gave little Shajira a naprapathic treatment, paying special attention to her upper and lower back. She winced and cried out when I applied pressure in order to restore proper circulation of blood and nerve currents. I hated to hear her cry but remembered the words of our professor at the College of Naprapathy: "Sometimes there is no gain without pain."

I tried to be as slow and gentle as I could, breaking the intensity of the discomfort with soothing massage to her neck and shoulders. She was obviously relieved when I finished.

I left her parents to dress her and asked them to wait for me on the veranda while I went into the forest behind our farm to search out the *Che Che* Xiv, the chief herb formula that Don Elijio had instructed me to use in cases of paralysis.

I found the *Eremuil* or Wild Coffee tree first and said the prayer of thanks to the spirit of the tree before taking the leaves. Next, the *Xiv Yak Tun Ich* or Pheasant Tail bobbed in the breeze as if to let me know where to find it quickly. It took a longer search to find the last ingredient of the mixture, the *Palo Verde* (Green Stick) leaves. Since it grows only near water, I had to climb down the steep riverbank and follow the shoreline to a rocky, eroded section where it grew.

The whole process took about thirty minutes, and the family was obviously relieved to see me return with a cotton sack full of the leaf mixture. I instructed them to boil a large double handful of the leaves in a five-gallon pot of water for ten minutes.

"Sit Shajira on a chair behind the steaming leaves and cover her and the pot with a warm blanket, leaving only her head exposed so that she can breathe," I told them.

Recently I had begun to take Don Elijio's warning about the Winds more seriously, especially in cases of paralysis and muscle spasms. So I added, "Close all the doors and windows, because if a *Viento*, Wind, should catch her during or after the steam bath she might get worse instead of better. Please, *mamá*, be very careful about that."

The mother shot me a look of surprise and understanding as if she didn't expect me to know about the dangers of *Vientos*, since most Americans consider it local superstition.

In the workshop, I poured off a pint of Jackass Bitters tincture from a gallon jug. I gave this to the parents.

"Give Shajira this wine tincture by the spoonful three times daily before meals. Tomorrow you are to give her whatever laxative you use for your children. What do you use?"

"My grandmother knows plenty herbs for that," answered the mother. "I will ask for her help."

"What is in the wine?" the father wanted to know.

"Do you know the plant called Jackass Bitters or *Tres Puntas?*" I asked. "It is excellent for viruses."

"Yes, I do know the plant," said the father with a smile.

"Give her lots of papaya juice, lemonade, and simple meals for fourteen days," I instructed them further. "Bring her back the day after tomorrow for another treatment. These herbs for the steam bath will last a week. Then I will pick them fresh for you again."

They nodded happily.

"Before you take her home, I would like to say the Maya prayers for her, with your permission," I said.

"Yes, of course," they answered in unison.

It was a moving moment for me. I felt her pulse and, by its rhythm, instantly knew which prayer to use. I repeated the prayer three times over her right wrist, three times over her left wrist, and three times over her forehead. As I looked down at her, cuddled comfortably in her mother's lap, our eyes met. She reminded me of a helpless, innocent creature one might encounter in the bush. I badly wanted to help this child and said an extra prayer for myself.

The family left in much better spirits than when they arrived. The father carried his daughter in his arms to the truck, where he placed her tenderly on her mother's lap. He covered Shajira with a blanket and rolled up the window. Good. Very smart, I thought.

Over the next few months, Shajira and her parents returned twice weekly for treatments, prayers, and herbs. For the first month, there was no improvement. But none of us lost faith. We consoled and encouraged each other and continued with the treatment program. I conferred several times with Don Elijio, who agreed completely with the course of treatment I had prescribed.

By the fifth week, Shajira was able to stand up unassisted. By the sixth week she could take halting steps, dragging her right leg. The treatments were still painful, but she was now comfortable enough with me so that I could manage to keep her laughing and hopeful.

Things really began to improve by the beginning of the third month. By now she could walk from the truck to the treatment room. The next week, she returned to school for the first time since her illness.

After twelve weeks, Shajira was 95 percent recovered. She still had residual tenderness in the spinal column, but her sacrum was no longer sensitive and she had regained full use of her legs.

I can't tell you how happy it made me to see her running and playing in the schoolyard.

CHAPTER TWENTY-ONE

Billy Webb Tree   Sweetia panamensis

The bark of the tree is boiled and drunk as a tea for diabetes,
tiredness, lack of appetite, delayed menstruation,
and dry coughs.

———————

Like a thousand other mornings, Don Elijio and I set out early in search of plants. This day our task was to locate Billy Webb trees from which we carefully strip long slivers of bark in a way that allows the tree to regenerate easily. The bark, boiled and drunk, was an important part of Don Elijio's arsenal of plants. He used it to treat diabetes and dry coughs, and to encourage the appetite.

The day before, we had finished our supply of *Zorillo* or Skunk Root. It seemed we could never keep enough of the foul-smelling root in stock. Many days we gave it to patients as soon as we set down our sacks, without a chance to chop and dry it.

A tea of *Zorillo* was used to cleanse internal organs and to help heal stomach ulcers. Baths in water steeped with the root helped many skin conditions common in the tropics. Don Elijio's nickname for *Zorillo* was *Metinche* (someone who puts their nose into everything), in reference to its versatility for both physical and spiritual ailments. Whenever he was confused about symptoms or didn't get the expected results with other herbs, he would prescribe *Zorillo*. So we needed also to restock our *Zorillo* or Skunk Root supply.

We had brought a light lunch of tortillas with us because the Billy Webb trees were a long, circuitous hike away from the forest footpath that lay about two miles north of the village.

Just walking to the trail took over an hour. "When I first started with this work, I could see my *farmacia* from my doorstep," he explained. That was the late 1930s. By the 1960s, he had to walk thirty minutes to reach the old growth forest where his vital remedies grew. In the 1980s, the healing forest was an hour's walk up a road, through merciless sunlight. The tall denizens of the roadside that had once provided shade for humans and shelter and food for wildlife and plants had vanished. The forest had given way to new fields, roads, and homes as the village had expanded and population had increased.

Still, it was a beautiful day and both of us were happy to be where we were, at home with each other and nature. Don Elijio was in a playful and happy mood, making jokes and telling stories all morning. We found the Billy Webb trees and said our prayers in thanks to the spirit of each tree before beginning. He carefully showed me how to use my machete to make oblong cuts in the trunk three feet above the Earth to prevent rain from splashing soil contaminated with bacteria into the incisions.

It takes a long time to skin a tree carefully, and it was almost midday before we were finished. We had stripped enough bark to fill a sack for me to carry weighing about fifty pounds. Then Don Elijio showed me other trees he had stripped before and how well they had healed. He caressed their new bark as if they were also his patients. They were.

We hiked further until we found enough *Zorillo* or Skunk Root to fill a sack for Don Elijio to carry—probably another fifty pounds. This he carried strapped to his head, Maya fashion. That method gave me a headache, so as usual I carried my load as a backpack strapped around my shoulders. That way the load rested on the small of my back, leaving my arms free to collect Xiv and to wield my machete.

It was 3 P.M. by the time we started down a steep hillside that led to the old logging road and back to San Antonio. Don Elijio was in marvelous form. In spite of his heavy load, he seemed to glide effortlessly down the slope, hardly catching a breath between stories. Above our heads was a playground of tropical birds and butterflies. I envied their weightless flight as I trudged along with my sacred burden.

Don Elijio suddenly slipped and fell, propelled forward by fifty pounds of Skunk Root. I gasped. But hardly missing a beat, he reached out and grabbed a sturdy vine. There he swung back and forth like a pendulum holding onto the vine with the sinewy muscles of his arms. He laughed out loud and said, "Ha! This is good exercise. I should fall more often."

"Don Elijio, you're a strong man," I told him once I had recovered from my momentary fright.

"Very strong," he answered, as he slithered down like a boy at play. "Enough blood and strength to keep a woman up all night long, kissing, whispering, kissing and whispering secrets. All night long. I don't tire."

Don Elijio looked so happy out in the forest that day, I wished we could stay there forever. As we resumed our journey downward, the heady, delicious aroma of humus enveloped us. Don Elijio stopped for a moment to adjust the straps around his brow. "When I was young I could spend the whole night in the forest," he said. "Would you have that courage?"

"No," I said right away. "At night the forest belongs to creatures like snakes and jaguars."

"What if we were to become creatures?" he asked. "I've never been tempted to become a jaguar, but, Rosita, I would if you and I could make a nest together in the forest."

I laughed and told him I was honored by the offer but I didn't want to be a jaguar any more than he did.

"Well, anytime you're ready I can pull that old prayer out of my head in a minute," he said, teasing me. We continued happily down the slope.

It was a few minutes later that we first began to smell the acrid, black smoke. The sky above the trees, once azure and white, was now ominously gray. Just ahead, our little footpath was blocked by flames licking their way hungrily into the forest.

We stopped in unison, shocked to see the abrupt change. Don Elijio cleared a side path that skirted around the advancing flames and motioned for me to follow him. I did as I was told, confident that he would lead me through the forest safely.

We made it down the hill, only to see that the field on the other side of the road was engulfed in flames. On the edge of the field, high swords of fire were consuming the cohune palms, fed by their abundant oil.

Farmers were burning their fields to reap the benefit of wood ash for fertilizer and to rid the soil of agricultural predators. But the *milpa* fires, intended to prepare the fields for crops of corn, beans, and pumpkins, were out of control.

"This fool has not made a firebreak," yelled Don Elijio. "No! No! This kind of farmer cares only for himself and damn his neighbor, the plants, and the creatures. They only fool themselves. Nature will make them pay for this cruel treatment. Soon, Rosita, there will be no place left for me to harvest God's medicines to heal his children.

"I've lived too long, that's what's wrong," he continued, standing agape at the sight before us. "I've lived beyond the plants and the forest and people who care. When they are gone, I would rather be dead too. What would there be for me to live for? Nothing."

"We have to get out of here, *papá*," I screamed at him, just as a ball of fire blew across the road and landed in the forest behind us. He started to leap for the flame to snuff it out, but I stopped him and did it myself. Balls of fire were flying everywhere now, fueled by sun-dried palm leaves and the afternoon breeze. More trees were catching on fire, and flames were spreading on both sides of the road.

I stamped and smothered what I could in a vain effort to prevent our medicine trail from being destroyed. Then we made our way along the road, following a trail of devastation.

The air was thick with smoke that burned our eyes and turned the already hot afternoon into a living inferno. Don Elijio's failing eyes became useless and he stumbled several times. Knowing he would never leave the bags of Billy Webb and *Zorillo* behind, I took his bag and staggered along, trying to carry both of our sacks. As I struggled with the bags, I was all too conscious of the fiery projectiles, falling trees, and my *maestro,* walking slowly in front of me, his shoulders slumped, dragging the pick, machetes, and hoe.

Several farmers had decided to burn together that day. The more careful, thoughtful men had made firebreaks to protect surrounding stands of forest. In these fields, the fire was contained and under control.

After getting through the worst of it, we stopped for a rest under a shady tree and shared an orange. Don Elijio sat down, defeated and exhausted, looking like the ninety-some-year-old man he was. His shoulders drooped and his eyes teared. He looked nothing like my playful companion who had swung on a vine over my head an hour ago.

We sat in silence. At last he said sadly, "That field we just passed was the last place anywhere around here where I could collect *Eremuil* leaves. A month ago, I went to the farmer and asked him to please spare that blessed tree because I need so much of it. Did he listen to me or my plea? No! Chopped it down and burned it all up! His own wife needs that tree.

"Now what, Rosita? Now what? Nothing does what *Eremuil* can do. It is the queen of all the Xiv. This is what comes from living too long. The day will come when there'll be no medicine left around the village and my art will be finished. Only tales and stories will remain. Where will my people get healing then? Where?

"I lost my daughter. I lost my wife. All of that I bore. But now I wish Saint Peter would find my name in his book and call me home."

It was hard to console him. What was I to say? I too felt discouraged and disheartened. Here was another swath of one of the world's last great rainforests going the same way as all the others. We never learn.

"There are still some *Eremuil* trees on my farm, *papá*," I told him. "I'll bring you leaves from those trees every week. I promise you'll never be without. Don't worry, please, my king. We'll help each other. Greg and I will go searching for your medicine wherever we have to go, we will."

At that moment I was struck with a plan. Why not talk to the farmers ahead of time and arrange for us to harvest their medicinal plants before they burned their fields and destroyed them? That's what Don Elijio's friend Don Antonio did for the plants on his farm. He harvested them and sold them before he burned so that less of nature's bounty would be wasted.

"What a shame the farmers didn't let us know they were burning today so we could have gotten some help and harvested the plants," I told Don Elijio. "Next year, we'll start asking in February before the March and April fires are set."

"Good idea," mumbled Don Elijio. "But what do I care? I'm dying and probably won't even be here next year. When I was young medicine was everywhere—easy to find and abundant. Now, ha! Harder and scarcer every year. Where will it end? This is a bad sign for me and my work. Worse for the people, though."

We picked up our burdens and made our way slowly back to the village. To the left were the charred remains of a piece of second growth forest. Tree stumps were still smoldering and the hilly landscape was gray, black, and barren. There were no signs of forest life anywhere, just the hot sun beating mercilessly down on the already-baked earth.

To our right was an untouched piece of woodland. The larger trees shaded our advance under cool breezes. A yellow flowering vine hung from a branch above. Butterflies romped, insects buzzed, and several species of rainbow-colored birds flitted and chirped in and out of the foliage. A chameleon darted for cover as we approached.

The contrast was sad and sobering. Don Elijio and I paused for a moment to contemplate the stark, smoking graveyard just across the road and what used to be and was no more. I felt as if my best friend, the forest, had a knife to her throat and I could do nothing.

That particular dry season, in 1989, was a low point in rainforest destruction in Belize. Never before had we seen such extensive burning of both small fields and large tracts for agriculture and community development. That year, too many farmers defied the rule to always cut a firebreak, and escaping *milpa* fires raged in every district, jumping over roads and fences. A black haze filled the air daily, and ashes fell everywhere.

When the rainy season came, this large-scale wrenching of trees from their deeply rooted beds caused the rivers to rise as never before. One large, mature tree alone can hold thousands of gallons of water in its trunk, leaves, and roots. But thousands of trees had been killed. Without trees, nature had no way to contain the mammoth amounts of rainwater that came pouring down the hillsides. Along with the water went the thin topsoil, turning the engorged rivers into churning mud that carried riverine plant and animal life.

The next year, several hundred acres of forest were cleared just miles upriver from our farm for development of a citrus plantation. One day over a period of five hours, the Macal River rose a record sixty feet, nearly arriving at our doorstep.

The destruction of the rainforest and its pharmacy seemed to drain away Don Elijio's spirit and stamina. He had been deeply shocked by that fiery day in the village fields. He spoke of it for a very long time, lamenting that he had outlived his plants and his friends, therefore his usefulness.

It wasn't long after that Don Elijio stopped going alone into the forest in search of his healing partners. By this time, the rainforest had receded even further—it was over an hour and a half walk from his door—and his eyesight had deteriorated to the point where he couldn't discern stumps and vines from creeks and rocks.

It became part of my service to him to be sure he was well supplied with his primary medicines of vines, roots, and barks as well as the Xiv (leaves) used for bathing and wound powders.

Since he required hundreds of pounds of dried medicines each week this was a formidable task. I had no idea where I was going to get enough plants to be able to supply both of us. Don Elijio knew the rainforest like no one else. He had roamed daily through fifty acres of high mountain government-owned forests above San Antonio, into places I was doubtful I could ever find alone. Some plants he purchased from old friends like Don Antonio who shared his respect for plants but were themselves rapidly succumbing to old age.

Some plants I purchased from herb vendors or paid people to collect. Greg and I and a local herbalist named Polo Romero also began visiting farmers before they burned their milpas, so that we could harvest their medicinal plants. Mostly, I spent hours alone in the forest and the fields, gathering his plants. Sometimes I collected on Ix Chel's thirty-five acres, or in the forests and fields near San Antonio.

Wherever I was, I missed my longtime companion. As much as I loved my solitary forays into the rainforest, it was never the same without Don Elijio.

CHAPTER TWENTY-TWO

Pheasant Tail   *Cola de Faisán*
Xiv Yak Tun Ich   *Anthurium schlechtendalii*

A bath or steam bath of these leaves boiled in water is considered
a specific treatment for muscle aches, backaches, rheumatism,
arthritis, paralysis, and swelling. The leaf is crushed with a stone
and then applied directly to sore muscles or backache.

———————

Antonio was a friendly vendor from whom I bought beets and onions
every Saturday morning. His booth was a regular stop on my weekly
shopping trip to San Ignacio's busy outdoor marketplace, overlooking
the Macal River.

One Saturday morning, Tony looked uncharacteristically grim.

"Are you still working with Don Elijio? Can you cure now, Doña
Rosita?"

I could tell from his expression that he was not just making small
talk. He seemed genuinely anguished.

"Maybe you could help me? My wife and I will come see you this
week, yes?" he continued. I was just about to say "of course" when he
added, "Things are not right for us, Doña Rosita. Something is not natural."

Those last few words pierced through the voices of buyers, shouting
orders for corn and plantains. I'd heard them enough times at Panti's
clinic and I'd been dreading the time when they would be addressed to
me. After seeing hundreds of cases at Don Elijio's clinic, I couldn't deny
the existence of evil forces. But it was one thing to assist Don Elijio. It
was another to make that final leap of faith alone into the frightening
and ambiguous sphere of spiritual healing.

Nor did I have confidence that I could treat spiritual diseases. Physical, yes, but spiritual? That was a different story. Was I as strong as Panti? I didn't think so.

"You better go see Don Elijio," I warbled.

"No, I want to see you, Doña Rosita. I have faith in you," he said.

I was unable to refuse his request. It was a genuine plea for help. I told Antonio to come to my farm as soon as he could.

Two days later, he and his wife Helena sat at my kitchen table, recounting their story. They lived in Benque Viejo del Carmen, where their grandparents and every family member since had been born. They had started their business by selling modest harvests off the back of their truck, eventually opening a permanent stand at the San Ignacio market.

Recently, they said, one of their neighbors had become jealous of their hard-earned success. The neighbor had been heard complaining that Antonio and Helena were too smug about their good fortune. He had told many people in town that he would put an end to their conceit.

One day while Antonio was away, a dish of stewed pork was sent to their home as a gift, delivered in the hands of a child. "My auntie sends this to you," the youngster told Helena. She didn't recognize the child and wasn't sure who had sent the plate of food, but she ate the stewed pork for lunch. By the time Antonio returned home from Belize City, she was strangely ill with nausea, acute anxiety, and an overpowering sense of doom.

Antonio called in the family's granny healer, who was respected by dint of her decades of experience and knowledge of medicinal plants. The crone immediately suspected the symptoms and asked Helena, "Did anyone send you something to eat?"

The grandmother concluded that the meat had been tainted with black magic by the jealous neighbor, whom she also had overheard boasting in public that he would teach Antonio and Helena a lesson about pride.

But Antonio and Helena didn't believe the old woman. They went to the town clinic, where Helena's sickness was diagnosed as gastritis. She took pills for a month but only got worse. The original symptoms intensified, and she had severe menstrual cramps for the first time in her life. She also began to suffer from nightmares, twisted fantasies, and depression and became indifferent to her roles as wife and mother.

Antonio brought Helena to Belize City to see another doctor. The doctor admitted that he didn't know what was ailing her but suggested it might be hysteria. From Belize City, they took a bus to Merida in Yucatán, Mexico, to consult with a specialist who took X rays and ran a series of tests with no conclusive results. He gave Helena Valium, which made her even more nauseated and depressed.

In utter desperation, having spent most of their meager savings, they again asked their granny healer for advice.

"You need to see a *curandero* who can do this work," she scolded. "Someone like Don Elijio in San Antonio. That is your only hope."

I had been wringing my hands throughout their story. I believed what they had told me. It was a classic *envidia* scenario. I felt sure the woman had been poisoned by the pork stew, which had been treated with enchanted powders and evil prayers.

Again I tried to send them to see Don Elijio, but they protested. "We have faith in you, Doña Rosita," they said.

My stomach was rolling and my tongue felt as thick as a foam pad as I reached for Helena's wrist and put my fingertips on her pulse. Instantly, I felt the sensation of painful stabbing soar up my hand and forearm. The familiar cold chills racked my flesh. Her pulse was extremely rapid, bouncing, and fat. It felt like firm, rushing bubbles under my touch.

As Jerónimo had done for Don Elijio, Don Elijio had done for me. For nine Fridays he had said special prayers that would protect me from the onslaught of evil forces while working with victims of black magic. Still, I was nervous and couldn't help wonder if the prayers would really work.

Then I looked into Helena's eyes. She gazed at me with intense confidence and hope. I fought back my reluctance, called on my faith, and let my healing instincts muscle their way past my trepidation and doubts. Without another moment's pause, I began reciting the prayers for *envidia* into her pulse.

I wished I could use my sastun to ask the Spirits to confirm my diagnosis of *envidia*. But the Spirits had never taught me to use my sastun as a divining stone. I could only use it to request dream visions and strengthen my healing powers.

Even if I had had a divining sastun, I wasn't sure if I would have been able to read it. Many times Panti had tried to teach me to read the little bubbles and black dots, but I had never been able to decipher their meanings.

I prepared dried Skunk Root, *Zorillo,* for her to boil into tea three times daily and instructed her to add the holy water from the Catholic church, Rue, and the sacred Esquipulas powder. Like Panti I prepared a mix of Copal and Rosemary for her to burn on nine Fridays.

By the time I had finished, Antonio and Helena looked remarkably relieved.

"Thank you, thank you, praise God that you are here to help us," they told me. "We were so weary of people telling us that there was nothing they could do for us. May God repay you and grant you a long life."

As they left, I wondered if my prayers would work. Then I remembered Panti's warning to never harbor doubt or the power to heal is severely diluted. "Don't say to yourself, 'I hope this works' or 'Maybe this will work.' No, you state clearly that you have faith with all your heart."

I reaffirmed my faith with a favorite prayer, "I believe, Lord, help Thou my unbelief," and went on with my day.

That night I had a dream vision. I saw myself lying on the grass in a field, facing the cloudless blue sky. I felt a sense of uneasiness and I was afraid, very afraid, without knowing why.

Suddenly, a powerful male Angel towered above my head. He wore the black and silver garments of a medieval knight and held a warrior's black shield and a menacing sword, which he seemed ready to brandish in my defense. Although his appearance was frightening, I knew he was there to protect me. I felt my fear drain away into the earth. He was the Spirit of *Zorillo.*

On my right was a magnificent, tall, female Angel, bedecked in a gown of sparkling golden hues. She arched her wings over me, enveloping me in her warm, amber luminescence. She was the Spirit of Copal.

To my left was a pure white female Angel, clothed in glowing crystalline garments that poured forth pearly white light. She stood serenely, emanating love and grace. She was the Spirit of Rue.

At my feet, another Angel stood. He was somewhat out of focus but I could feel his presence. He too was a powerful guardian. He was the Spirit of the stone of Esquipulas.

The Spirit of *Zorillo* was fiercely protective, reminding me of a samurai, ready to pounce. He was on guard, scanning the distance for any sign of evil or danger. The other Spirits—Copal, Rue, and the stone of Esquipulas—were equally powerful but not warriors. They emanated strength and warmth and a supreme confidence in their abilities to protect me.

I was engulfed by their protective shields and knew no harm could penetrate.

I woke up feeling loved, sheltered, and fearless, knowing I had a team of Angels by my side forevermore.

Two weeks later I saw Antonio and Helena in San Ignacio selling watermelons and okra off the back of their truck. They spotted me and rushed over to embrace me with exclamations of love and gratitude.

"I am completely back to my old self," said Helena joyously.

"You have lifted us up," added Antonio. "There is no one like you and God, Doña Rosita."

CHAPTER TWENTY-THREE

Cross Vine   Cruxi   Paullinia tomentosa

A common, weedy vine whose leaves make up part of the Nine
Xiv formula of herbal baths to treat any illness.

I woke up early the next morning, my forty-seventh birthday. I hadn't been to see Don Elijio since healing Helena, and I was anxious to tell him about what had happened.

Dawn was just breaking as I crossed the river and hiked to San Antonio in record time. This day my thoughts were of the last four years I had spent as Don Elijio's apprentice. So much had happened. As I walked I was joyfully aware of being able to recognize dozens of medicinal plants and trees along the roadside.

When I arrived I found him sitting quietly alone in the doorway scraping the Bayal vines he used to weave the *Escoba* palm leaf brooms.

"I cured a case of *envidia, maestro,*" I told him excitedly, still amazed. "I did it just the way you taught me and it worked. The woman is completely recovered."

He looked unimpressed and launched into his usual sermon about envy and greed. After all, he seemed to say, it wasn't a miracle, just the fruits of his teachings and the expected benefits of the medicinal plants and prayers.

Then I told him about my dream. That impressed him, and he wanted to hear every detail. "I told you they were good friends," he exclaimed. "Now you know you can do this kind of work with their help and protection. We are not alone."

We finished the brooms together, piled them into a corner, and spread out the canvas cloths and plastic sacks we used for chopping. He sharpened our machetes and pointed to a gnarled pile of Man Vine. He seemed uncharacteristically quiet.

He settled down on his seat across from me on the floor and grabbed a twisted mass of sweet-smelling vines to pull them apart.

"There is nothing more I can teach you, daughter," he announced. "You have learned everything I know and can do everything I can do."

I stared at him. "Oh no, Don Elijio, that's not true," I exclaimed. "Every time I visit you I learn something new, something I feel I never could have put together on my own."

He shook his head and continued, "I know it to be so. I've listened to you take care of my patients. You have even become my doctor. Never once have I felt that I had to correct you or that you gave bad advice. You have a pure heart. This is your calling in life, just as it is mine."

"But I still haven't learned to read the sastun," I reminded him. "I don't think I have the *don,* the lamp to see what the bubbles mean."

He reminded me that reading the sastun was not the only tool at my disposal. "You knew that the woman had envy from what she told you, and it will always be that way. These things are no great mystery.

"A healer must accept his strengths and weaknesses, Rosita," he added. "The most important thing is that the Spirits are with you. They see you and that you are working, and they will look after your needs. Long after I am gone, they will be here. Have faith, my daughter, for with faith everything is possible."

I didn't know whether I couldn't read the sastun out of fear or because I simply did not possess the psychic gift. I did know that my true love was plants and my *don* was that my hands could see through flesh and tissue. Of this I had no doubt.

I also had the gift of faith. I had grown to love and feel comfortable with the Maya Spirits and felt as if they loved me in return. I sensed their loneliness and their affection for me. I remembered that Don Elijio had been concerned that they might not communicate with a *gringa.* But they had—even speaking English, the language of my dreams.

Perhaps, it was possible that one day I would be able to read the sastun. "*Poco a poco, paso a paso,*" he always said.

A village woman came in with one of Don Elijio's many goddaughters, asking for a belly massage. He climbed to his feet with a groan and led her into the examining room. The child stayed with me, and we played the Maya nine-stone game that always reminded me of jacks.

The woman and child left, and from the doorstep of the cement house Don Elijio called me.

"Come, Rosita," he signaled.

I followed him in, and he sat down in his customary seat.

"This is for you," he said, holding out his hand. The small shiny marble rolled along the crevices of his palm. It was his sastun.

"What?" I asked, taken aback. I was shocked. I waved his hand away.

"I am ninety-three now and I'm dying soon," he continued, matter-of-factly.

"Don't talk that way, *papá*," I cried. "I hate it when you do that. Sometimes I think you could outlive me, you have so much energy and strength. What would you do without your sastun?"

He reached into his bag and pulled out another bundle swathed in cloth. "I've been given another," he announced. He unwrapped it and showed me a stone that was larger and paler than the one that he had used for sixty years.

"Last week, Rosita, I had a dream vision," he told me. The same old Maya Spirit in ancient ceremonial garb, who had heralded his first sastun, had returned.

"The old Maya said, 'We see that you are working hard, that you are old and tired and need some help. It is time for you to have a new sastun. In the morning at the first light of dawn, open your door and look on your doorstep. There you will find a gift to help you.'

"When I woke and heard the rooster crow and saw the light through the crack of the window, I jumped up and opened the door," recounted Don Elijio. "There on the doorstep sat this new sastun.

"I want you to have my old sastun because you need it," he said. "Even though you can't read it yet, you can use it to enchant *protecciones* and photographs."

He dropped his old sastun into my hand. It felt cool and light. I accepted and told him that today was my birthday.

"*Bién suave,*" he said, grinning. How smooth.

Two mornings later, Don Elijio didn't get up at his usual hour of dawn. At dinner the night before he had complained of stomach cramps. During the night I had heard the ropes of his hammock creak, as he shifted and sighed, unable to sleep.

Shortly after sunrise, I heard him gasp, a long painful breath. I parted the curtain, rushed over, and pulled his blanket back.

"Are you all right, *papá?*" I asked.

"Ahh, Rosita," he said in between gulps for air. "I'm dying."

"What's wrong?" I asked.

"I have had this terrible pain in my belly all night," he whispered. "I feel like there is a tiger in my guts. The pain is unbearable."

I took his pulse, said the prayers for *ciro,* and massaged his abdomen. Then I ran into the kitchen hut, started a fire, scooped out a gourd of Man Vine from a sack, and put it on to boil.

"I am on death's doorstep now," Don Elijio wailed as I spooned the tea between his dry, rubbery lips. "I can see Saint Peter beckoning to me, calling me home. Last week I saw Chinda in my dreams, Rosita, for the first time since she died. She looked so fat and well. She told me I looked pale and thin. She whispered, 'I'll come to get you soon, sweetheart. Not long to go now.' I got up from the hammock to embrace her, but she said, 'No, not yet,' and disappeared.

"I begged her to return but she didn't," he said as he clutched his belly.

Now that I had done what I learned from him, I decided to try an additional therapy that I had found worked well in cases of stomach cramps. I heated castor oil in a small clay pot over the fire and soaked a cotton cloth in the warm oil. Then I put the cloth on his abdomen and placed a hot water bottle that I had once given him over it.

The castor oil pack was to be on for an hour, so I sat on a stool next to the hammock.

He was very depressed. "What's wrong with me is old age and loneliness," he moaned pitifully. "Where I find the cure for this is six feet underground. Has Saint Peter forgotten to call my name? What is an old man to do? I am ready to die."

He told me he wasn't afraid of dying but he was worried about his sins. Had he done enough goodness to make up for his transgressions?

"What do you mean, *papá?*" I asked, stroking his forehead.

"I was a drunk and I know that made Chinda suffer," he cried. "My horse would come home alone, and she had to ride out to find me sleeping in a ditch. All alone she would pick me up, put me on the horse, carry me home, put me to bed, and make teas for my hangover."

"But you stopped drinking," I said.

"Not until after poor Chinda died," he said.

"We are all capable of sinning, *papá,*" I said. "I know that God sees the good you have done. Just think of the thousands you have lifted up."

I saw that he was crying and bent over him. He seemed so frail and small that the hammock nearly swallowed him up.

"I slept with many women," he whispered softly. "Chinda never knew. But I never sinned with a patient. And I never used my sastun to enchant a woman for myself. I swear on the souls of my great-grandchildren."

I was a little surprised but not really disappointed. He was a Latin male who had been taught to live the *machismo* code, and women of Chinda's generation had accepted their husbands' indiscretions so long as they were loved. Chinda had been loved and cared for as few others, of that I was sure.

I believed him when he said that he had never used his sacred powers to enchant a woman for himself, knowing that a H'men is forbidden to use his own powers for personal gain. He hadn't enchanted any widows or even La Cobanera, preferring instead to suffer in loneliness. And he had never enchanted me. He was an incurable flirt, but he had always respected the boundary between friendship and romance.

"*Papá,* loving women is not the worst sin," I told him. He was sobbing openly and clutching my hands. I held him and tried to soothe him.

"Always it is only me in the hammock with no one to warm my old bones or whisper secrets in my ear. It is painful but I deserve it."

I couldn't bear to watch him mourn his life as if it were a charred slate of sin and deprivation.

"But, *papá,* you forget the thousands of people you've lifted up," I cried. "Surely God knows you're a human man. He knows what you've done on this Earth."

Now I was crying, desperate to ease his pain. He had often told me: get patients to laugh and half their troubles disappear. It was still some of his best advice.

I searched my mind for a joke to tell him. The only one I could think of was a little dirty but seemed appropriate.

"I have a *chiste* for you," I told him. Despite his misery, I noticed a flicker of interest.

"There were once two twin brothers who were very close," I told him. "One was very good and pious, and the other was a drinker and a womanizer. They died together in a car accident. One went to heaven and one went to hell. The good brother spent his days sitting on a cloud listening to heavenly music. One day he got permission to go visit his brother in hell. There he found his brother in a saloon with a bottle of beer in his hand and a woman on his lap, having a grand time.

"The good brother went back to Saint Peter and complained. 'He's having a great time in hell while my life is boring, sitting on a cloud and doing nothing,' he said.

"'Ahhhhhh, don't worry about that,' Saint Peter said. 'The bottle has a hole in it and the woman doesn't.'"

Don Elijio roared. The hammock shook. Over and over he kept repeating the last line of the joke and giggling.

I noticed some color creeping back into his cheeks. It was time to get my loquacious friend talking again.

"Greg and I have decided to do a *Primicia* every month," I told him.

"Ahhh, that's good, daughter. The Maya Spirits are almost as lonely as I am," he chirped as I adjusted the castor oil pack. I rubbed his feet, which always soothed him tremendously.

"I remember when you first came around to see me," he told me. "My relatives told me not to trust you. They said that your interest in me was not good. They were wrong. Through all these years you have been my friend. Friendship is what counts. Now there is only you to carry on my ways. You have given me as much as I have given you."

Tears welled up in my eyes as I leaned forward and vowed, "*Papasito,* I will be with you until the last step. I will never leave you."

I sent a message home with Angel and stayed with Don Elijio for three more days, monitoring his condition and taking care of all his patients. I continued to treat him with prayer, massage, and the castor oil packs.

On the afternoon of the third day he was sleeping in his hammock while I sat on his customary stool talking to a heavy-set, middle-aged East Indian woman who had come to consult with him.

"I am sorry," I told her. "Don Elijio can't see patients today. He has been very ill and is resting now. Either I can help you or you'll have to come back another time."

But the woman was insistent and kept peering behind the curtain to where he slept.

Just as she was about to reluctantly resign herself to my services Don Elijio came staggering out in his underwear.

"*Mamasita! Mamasita!* I nearly died!" he shouted, gesturing wildly. "I was as close to dying as I ever came in my life. I got right up to the gates and met Saint Peter. He looked at me and said, 'Where have you been, old man, get in here. Someone must have forgotten you.'

"'Now just a minute,' I told him, 'let me ask you something before I go in. Is there beer? Are there women? Is there dancing?' Saint Peter said, 'Beer, women, and dancing? Are you crazy? There's none of that up here.' I answered, 'Forget it, I'm not going in.'"

Panti threw up his arms and chanted, "I'm back!"

"You better go back to bed," I told him.

"No, no, let's work. What does this lady want? Wait till I put my pants on."

He scampered back into the bedroom and reappeared moments later.

It was clear he knew exactly why he was put here on Earth.

The woman gave me a dirty look.

I got up. Panti sat down in his chair and launched into a new routine.

"What is your problem, *mamasita*? I'm 101 years old, and whatever ails you I can cure with my prayers and my herbs and God at my side."

I shot him a look of surprise, wondering how he had managed to age eight years in a matter of minutes. He ignored me.

"I cure diabetes, high blood pressure, arthritis, broken hearts," he continued. "I've been doing this work for forty years and I know a few things."

I smiled, remembering that I had once asked him why he always used that number, since by my calculations he'd been practicing bush medicine for far more than sixty years. He'd just shrugged and said, "Because, child, that's as high as I can count."

After that day Don Elijio continued to practice for many more years, and he is still there today. On days when he feels good, he sees the patients who always seem to find their way to his door. Other days, when his body aches too much, he puts up a cardboard sign that Angel made for him. "*Cerrado,*" it says. Closed.

Don Elijio is now 97, still looking for a wife and telling his patients that he is 101. God willing, he will live far beyond that.

EPILOGUE

Man Vine   Behuco de Hombre   Ya Ax Ak
Agonandra sp.

A forest vine whose woody tendrils are chopped, boiled, and
drunk for all manner of ailments of the digestive and eliminative
tract. Both the root and vine are used for male potency, the root
being the stronger of the two for this purpose. One of the
primary remedies in Maya medicine.

In the fifth year of my apprenticeship, I sat down at my old typewriter
and pounded out letters to hundreds of scientists around the world. I felt
that I had a responsibility to properly record Don Elijio's work for pos-
terity. I asked them for help, guidance, and suggestions.

Most never responded. A few wrote back saying, "What a fascinat-
ing project. We have no funds or time to assist you, but good luck and
keep up the good work." It was discouraging, but I didn't give up.

One day, a tourist at Chaa Creek gave me a copy of a newspaper ar-
ticle about the worldwide search for medicinal plants with anticancer
and anti-AIDS compounds. The scientist in charge of these plant collections in Central America was Dr. Michael J. Balick, director of the
New York Botanical Garden Institute of Economic Botany in the
Bronx, New York. I wrote him immediately.

I was surprised to receive an enthusiastic letter in response exactly
one month later. Dr. Balick was more than just interested. He wanted to
come down to Belize to meet me and Don Elijio. True to his word, he
showed up fifteen days later.

As soon as I saw him, I liked him. He was handsome and tall, a little
on the chubby side, with intelligent, jovial eyes and a good sense of

humor. He was humble and willing to learn. I knew right away that he was for us and we were for him.

Over dinner that night on the farm, he told us about ethnobotany. I had never heard of the science and hadn't realized that my work with Don Elijio would be considered ethnobotany. But it was. *Ethno* means people and *botany* means plants. Ethnobotany, then, is the study of how people and plants interact.

The next morning Dr. Balick and I walked the five miles to San Antonio together to meet Don Elijio. They liked each other immediately, and before long the Harvard-educated scientist and the gifted H'men were sharing jokes, laughing and entertaining each other in Spanish.

We spent the day in the clinic, then slept in the waiting room hammocks. The next morning, when the light broke, the three of us walked the rainforest path and went in search of *Zorillo*.

I could tell that Dr. Balick was as taken with Don Elijio as I was, and I wasn't surprised when at the end of the day, he leaned over and said, "Well, Rosita, there's no doubt that this man is an authentic Maya traditional healer whose knowledge must be recorded."

Dr. Balick, by now Mike, wanted to return to Belize to collect plant specimens. He explained to Don Elijio that his mission was to find medicinal plants that might help in the search for a cure for cancer and AIDS, then asked the old H'men if he would help.

"So you scientists aren't so dumb after all," quipped Don Elijio happily, slapping his knee. He looked extremely pleased and agreed to share his knowledge.

A few months later, Mike returned. Greg and I accompanied him out into the field and worked sixteen hours a day for thirty days collecting two hundred plants. In the process, Greg and I learned how to conduct proper field ethnobotany. We brought the plants to Don Elijio, who explained how he used each one, then prepared and shipped them to the National Cancer Institute laboratory in Washington, D.C.

Mike had the idea of asking Don Elijio to choose his twenty-five most important medicinal plants. The laboratory extracted and processed those twenty-five plants before the rest. The researchers were thrilled at the results. Usually 1 to 5 percent of all collections show activity in the lab. Twenty-five percent of Don Elijio's twenty-five plants showed activity, which was considered phenomenal. The results were written up in different journals across the United States.

We were very encouraged. This rare collaboration between a traditional healer, a scientist, and an alternative physician had allowed us to pool our knowledge and resources, achieving spectacular results. By focusing on Don Elijio's favorite plants, we had reduced some of the randomness of scientific study.

We then created the Ix Chel Tropical Research Foundation, fulfilling our dream of creating a bridge between science and traditional healing for the benefit of humankind. The foundation has sent over two thousand plants to the National Cancer Institute. Of these, the NCI has found twelve particularly promising plants that it continues to study.

Five hundred of those two thousand plants came from Don Elijio's memory. Other plants were brought to our attention by other traditional healers in Belize. Among the most well known and respected were Hortense Robinson, herbal midwife of Ladyville; Polo Romero, herb gatherer and snake doctor of San Ignacio; Andrew Rancharon, snake doctor of Ranchito Corozal; Juana Xix, granny healer and midwife of Succotz Village; and Thomas Green, our very own "Old Man River" who ran the dory and turned out to be a very knowledgeable healer.

It was my role to locate the traditional healers, explain our mission, and ask if they would be willing to show us their plants and share how they used them.

What a joy! It was a job I cherished. I have found that traditional healers tend to be highly intelligent, admirably self-confident, deeply spiritual, and humorous. They care more for the welfare of others than they do about themselves.

As I talked with traditional healers, I realized that they had much in common but little contact with each other. Some of them didn't even know the healers in other parts of the country. So Greg and I came up with the idea of setting up traditional healers conferences in Belize. Funded by the United States Agency for International Development and assisted by the Belize Center for Environmental Studies, we held five countrywide conferences. Hundreds of healers and lay people attended. Many of the traditional healers I met there have become my closest friends, enriching my life immeasurably.

Out of these meetings grew the Belize Association of Traditional Healers, BATH, an organization dedicated to furthering and enhancing traditional healing. BATH is also meant to be a conduit for shared profits of any drug developed from a medicinal plant collected in Belize. We hope that this will encourage pharmaceutical companies to share their profits with the people from whose knowledge their drugs are derived.

BATH has worked closely with the Ix Chel Tropical Research Foundation to educate people about medicinal plants. Probably our biggest educational tool is the Panti Maya Medicine Trail, named after Don Elijio. In 1987 my son James and I bushed a trail through the jungle behind our farm. James painted signs for thirty-five of the most interesting, useful plants found growing wild on the trail. Greg built a healer's hut—an exact replica of Don Elijio's old thatched clinic. At first just a handful of people found their way to the trail, but today thousands of Belizean college students, schoolchildren, Girl Scouts, tourists, scientists, and health professionals regularly visit the trail and healer's hut museum, the first of their kind in Central America.

We spent five years developing a trail guidebook with illustrations. This led to a coloring book for children called *Useful Plants of the Mundo Maya,* starring Ix Chel. Mike Balick and I also published a book called *Rainforest Remedies: 100 Healing Herbs of Belize* (Lotus Press).

Greg and I started offering seminars attended by students of University College of Belize, the Pharmacy School, the School of Nursing, and international health professionals and lay people. I began teaching a course entitled "Medicinal Plants" at the Belize College of Agriculture in Cayo District. Sponsored by the United Nations High Commission for Refugees, Greg and I also conducted workshops for community health care volunteers, teaching them how to integrate safe, reliable herbal remedies into their village clinics. We also collaborated with UNHCR to create an educational video entitled "Bush Medicine: A Belizean Tradition."

Since the terrible fires of 1989, we had been harvesting medicinal plants from small farmers before they slashed and burned their land. It was a huge job. We collected the plants by hand, then chopped, dried, and stored them. Some we gave to Don Elijio; others we packaged for use in our practice or to sell to visitors on the Panti Trail.

One day this project grew to proportions we had never imagined. We heard about a 1,400-acre tract of old growth rainforest called Beaver Dam on the Western Highway, which was being developed by Fortune International of Seattle. We were horrified at the thought of such massive destruction. There was no way to stop the development, but after some negotiations, we were given permission to harvest the medicinal plants.

Faced with the daunting task of collecting enough medicinal plants to fill a warehouse, Greg and I realized we had to have a more efficient method and a bigger market. Thus we created a company called Rainforest Remedies, which shares profits with employees and traditional healers. We developed fourteen formulas that use the plants most likely to be found in forests that we could harvest just ahead of fires and bulldozers. The formulas include Kidney Tonic, Traveler's Tonic, Immune Boost, Belly Be Good, Flu Away, Male Tonic, Female Tonic, and Blood Tonic.

The formulas are tinctures, which retain potency for years, therefore eliminating the need for drying and storing bulky bags of plants. As a result, we can process thousands of pounds of medicinal plants when necessary.

Rainforest Remedies products are marketed in Belize, the Caribbean, and Europe. Due to stringent Food and Drug Administration regulations on the labeling of herbal products, we cannot sell them in the United States, other than on an individual basis.

Our research center has also branched out into other areas. We are working with Anabel Ford, an archaeologist and professor at the University of California at Santa Barbara, who is excavating to help develop the ruins of Pilar, near the village of Bullet Tree Falls in Cayo District, a joint archaeological–medicinal plant park. The park features many of the plants that would have been used by the ancient Maya who once lived at Pilar.

One night, at a farewell party for Dr. Ford and her team, our friend Daniel Silva, the Belize Minister of Trade and Commerce, remarked, "You know, Rosita, we should develop a medicinal plant reserve in Belize."

"What a great idea," I said, "I'd love to help." A few days later, the minister gave us a map of the Yalbac Hills, about ninety minutes by car north of San Ignacio near the border of Orange Walk District and the refugee village of La Gracia. We drove out there and found an enchanting tract of old growth forest replete with rare medicinal trees, roots, vines, and herbs. It was also the refuge of jaguars, monkeys, peccary, deer, and curassow birds.

In June 1993, then Deputy Prime Minister Florecio Marin signed into law the development of a 6,000-acre medicinal plant reserve, perhaps the first of its kind in the world. We named it the Terra Nova Medicinal Plant Reserve.

The purpose of Terra Nova is to guarantee that future generations of healers will have medicinal plants to harvest. It also provides a home for the seedlings our teams rescue from areas of rainforest development. They'll be transplanted to Terra Nova so that they may live and bear fruit from generation to generation. It will be a place where people can come from all over the world to see rainforest plants thriving undisturbed in their own habitat. We believe strongly that the time has come to set up such reserves all around the globe.

The Belize Association of Traditional Healers will manage and operate the reserve, which is currently funded by private donors and grants. Eventually we hope that Terra Nova will be self-supporting. Greg and I plan to relocate and donate the Panti Maya Medicine Trail, now a popular tourist destination, to Terra Nova. The trail's profits will support projects, jobs, education, and the development of the reserve. There will also be a factory where refugee women from La Gracia Village will make products for Rainforest Remedies, thus providing income sources for families who live near the reserve.

Meanwhile, Ix Chel Farm has flourished. Thick grass grows over the five acres that we cleared. The trees bear tropical fruit year-round, and we always have plenty of organic vegetables to cook up for dinner. The farm is no longer the isolated place it once was. Thanks to a new all-weather road built by the British Army Corps of Engineers, Ix Chel has become a meeting place for scientists and guests. We also see our patients on the farm, having long ago closed our clinic in San Ignacio.

People know that if they can't see Don Elijio, they can come and find us at the farm. Although it has not always been easy, I've kept the promise I made that day in the cornfield to help, to learn, to heal.

Back at San Antonio, Don Elijio still sees patients when he feels well enough. The clinic is the same, although both of the old huts have died of natural causes and a single other, built by a pleased patient, stands in their place.

There are a few other changes worth reporting. Don Elijio now has three young Maya apprentices: his great-grandson, his grand-niece, and Don Antonio Cuc's grandson. And on the wall inside, across from the patient chairs and next to the Coca-Cola calendar, hang two new framed certificates and one plaque, all in English.

One is from the University College of Belize, awarding Don Elijio a special certificate for having greatly contributed to the development of the country. The second is from the New York Botanical Garden in appreciation for his participation in the search for plants with anticancer and anti-AIDS activity. The plaque is from Help Age Belize, declaring my *maestro* "A Most Valuable Senior Citizen."

"These are my guarantees," he tells patients proudly. "One is from the queen of England and one from the president of the United States.

"I used to be only a bush doctor, but now I am Dr. Panti—and I've been healing in this way with my plants and my prayers for forty years. This is my gift, my *don*. I never went to school, but up here, it's full."

c'ox ca'ax: "Let's go to the mountains."
comal: a round clay disk used to make corn tortillas and to toast food or plants
H'men: Doctor priest(ess), "One who knows"
ik: wind
inca: "I'm going now."
sas: light, mirror
sastun: a divining stone used by the H'men
tato: an affectionate term used to address the elderly and the revered
tun: stone, age
Tzibche: (literally) letter tree

A BASIC CATALOG OF
MEDICINAL RAINFOREST PLANTS

Allspice
Pimienta Gorda
Pimenta dioica

Use the leaves and berries as an old house-hold remedy for stomachache, colic, indigestion, and fever. Apply the crushed berry over the gum of an aching tooth to bring quick relief.

Amaranth
Amaranto
Calalu
Amaranthus sp.

A "pot herb" used for both food and medicine. Leaves are high in iron and calcium; seeds are rich in protein.

Anal
Psychotria acuminata

Several varieties; all are used for part of herbal bath formula.

Avocado Pear
Aguacate
On
Persea americana

Boil leaves with other species for cough syrup. Drink as a tea for pain, colds, and fever.

Balsam Tree
Balsamo
Na Ba
Myroxylon balsamum

Boil and drink bark for all conditions of urinary tract, prostate, liver, and inflammation.

Basil
Albahacar
Ca Cal Tun
Ocimum basilieum

A wild and cultivated spice also prized for its use to ward off evil spirits and break spells.

Billy Webb Tree
Sweetia panamensis

Boil bark and drink for diabetes, uterine congestion, to cleanse internal organs, for low appetite, and for dry coughs.

Breadnut
Ramon
Chacox
Brosimum alicastrum

A major source of staple food to the ancient Maya, and still eaten today. Boil the nut to make gruel, cook like a tortilla or like new potatoes. Feed the leaves to horses, cows, sheep, and goats with recent young to increase milk supply.

Cancer Herb, Cat Tail
Hierba del Cancer
Acalypha arvensis (male)
Priva lappulacea (female)

Boil entire plant to bathe chronic skin conditions and ulcers. Mix with *Tres Puntas* (Jackass Bitters) to make a dry wound powder. Drink tea of leaves for stomach ailments or ulcers. Best to combine male and female if possible.

Castor Bean Tree, Oil Nut
Ricino, Iguerra
S'Kotch
Ricinus communis

A weedy tree, all parts of which are toxic. Oil is a purgative when taken internally. Apply oil locally to heal aches, pains, swellings, and bruises. Apply heated leaf to skin for the same purposes.

Chamomile
Manzanilla
Anthemis nobilis

Used universally as a mild tea for nerves, colic, sleeplessness, and indigestion.

Chaya
Jatropha aconitifolia

A semiwild backyard leaf vegetable high in iron and minerals. Eaten much like spinach.

Chicle Tree
Sapodilla
Zapote, Ya
Manilkara bidentata

Principally valued for its resin, used as a base for chewing gum until the 1930s. Has a delicious fruit much prized by ancient Maya.

Chicoloro
Strychnos panamensis

Boil vine and drink for constipation, to cleanse internal organs; use as uterine lavage and as an antidote to poisons.

Cilantro *Coriandrum sativum*	A cultivated backyard vegetable used as a flavoring in sauces, soups, and salads. Make a tea of boiled seeds for insomnia and indigestion.
Cockspur *Zubin* *Acacia cornigera*	Bark is male aphrodisiac; thorns and their resident ants are used to loosen mucus in infants.
Copal Tree *Pom* *Protium copal*	Burn dried resin as incense to ward off evil spirits, witchcraft, and spiritual diseases such as envy, fright, and grief. Bark may be boiled and drunk for stomach upsets and intestinal parasites.
Corn *Maize* *Im Che* *Zea mays*	Sacred food to the Maya. Boil corn silk hair as a tea for all problems of kidney and bladder. Useful in preventing bed-wetting.
Cotton *Algodon* *Tsiin Taman* *Gossypium hirsutum*	A preconquest plant. Boil leaves in sugar with other species to make cough syrup.
Cow's Hoof Vine *Pata de Vaca* *Ki Bix* *Bauhinia herrerae*	Boil and drink vine to staunch bleeding and hemorrhaging, and take during menses as a birth control agent.
Cross Vine, Skipping Rope *Cruxi* *Paullinia* sp.	Boil leaves to bathe skin conditions, headaches, insomnia, and diseases of childhood.
Duck Flower *Contribo* *Aristolochia trilobata*	Soak vine in pure water all day and take one-fourth glass three times daily for sinus congestion. Plant has some degree of toxicity. Boil vine and drink for fever, gastritis, high blood pressure, and to cleanse internal organs and urinary tract of phlegm.

Fiddlewood, Walking Lady
Yax Nik
Vitex gaumeri

Dry and powder white inner bark to sprinkle on bay sore (leishmaniasis).

Give and Take Palm
Escoba
Crysophila argentea

A thorny trunk palm with many local uses. The dried leaves are woven into brooms. Scrape the inner bark and apply it to wounds and cuts to stop bleeding. The heart of this palm is edible.

Guaco
Aristolochia odoratissima

A woody vine used as a tea for all manner of gastric complaints.

Jackass Bitters
Tres Puntas, Mano de Lagarto
Kayabim
Neurolaena lobata

Boil leaves and drink for malaria, ringworm, intestinal parasites, amoebas, fungus, delayed menses. Dry and powder leaves to sprinkle on stubborn wounds and skin ulcers.

Lemon Grass, Fever Grass
Zacote Limón
Cymbopogon citratus

A pleasant-tasting herb used as a beverage and to reduce fevers in children and adults.

Linden Flowers, Basswood
Flor de Tilo
Tilia cordata

A mild sedative tea.

Mango Tree
Mangifera indica

Boil leaves with sugar until syrupy for cough, or boil with water as a tea to relieve menstrual cramps, headaches.

Man Vine
Behuco de Hombre
Ya Ax Ak
Agonandra sp.

Boil and drink vine for gastritis, constipation, indigestion, nerves, fever, and muscle spasms, and to cleanse internal organs. Boil and drink root for male impotency.

Mexican Wormseed
Epasote
Chenopodium ambrosioides

Take juice of fresh plant for intestinal parasites. Drink tea of root for hangovers. Add leaves to bean pot to prevent flatulence.

Naked Indian, Gumbolimbo *Palo de Turista* *Chaca* *Bursera simaruba*	The bark is a natural antidote for poison-wood condition, used to reduce fevers or as a bath for skin conditions, burns, blisters, bites, rashes, measles, and infections. Drink for kidney infection, stoppage of urine, dropsy.
Palo Verde *Eupatorium (Critonia) morifolium*	Boil fresh leaves for use as herbal bath for any ailment. Part of the Xiv formula.
Pheasant Tail *Cola de Faisán* *Xiv Yak Tun Ich* *Anthurium schlechtendalii*	Boil leaves and use as steam bath for rheumatism, arthritis, swellings, paralysis, and fluid retention.
Rosemary *Romero* *Rosemarinus officinalis*	Drink tea of leaves to cleanse stomach of mucus. Burn with Copal resin as incense to ward off evil spirits and envy.
Rue *Ruda* *Sink In* *Ruta graveolens*	Squeeze fresh plant into water and drink for hysteria, menstrual cramps, stomachache, and onset of epilepsy. Take with *Zorillo* and white stone to ward off spiritual diseases such as evil, envy, or fright. Fresh plant also used for delayed menses, labor, and delivery.
Skunk Root *Zorillo* *Payche* *Chiococca alba*	Boil and drink root or bark of vine to dispel envy and evil spirits. Drunk by shamans to increase powers, used to cleanse internal organs for stomach ulcers, and as a bath for many skin conditions or sores on mucus membranes.
Sour Orange Tree *Naranja Agria* *Citrus aurantium*	A wild tree whose fruits are used as substitutes for lemons. Drink an infusion of the leaves for fever, colds, flu, and hangover.
Soursop *Guanabana* *Annona muricata*	Boil leaves in sugar with other species to make cough syrup. The fruit is commonly made into ice cream.

Spanish Elder, Buttonwood *Cordonsillo* *Ixu Bal* *Piper amalago*	Over twelve varieties found in Belize; all are medicinal. Mostly used for herbal bathing for a variety of ailments, especially for skin conditions, headaches, nervousness, insomnia, and children's disorders. Root of most varieties is chewed for toothache.
Tzibche *Crotolaria cajanifolia*	Used in herbal bath formulas and to brush *Primicia* participants to protect them from being harmed by the Winds of the Spirits.
Vegetable Pear *Chayote* *Cho Cho, Wiskil* *Sechium edule*	A backyard vining vegetable bearing pear-shaped fruits with a mild flavor reminiscent of zucchini. Drink mixture of fruits and leaves boiled in water for high blood pressure and high cholesterol.
Wild Coffee, Suprecayo *Café Sylvestre* *Eremuil* *Malmea depressa*	Prime ingredient in herbal bath formula. Use alone as bath for stubborn conditions, especially for backache, muscle spasms, hysteria, nightmares, and insomnia.
Wild Poinsettia *Flor de Pasqua Sylvestre* *Euphorbia pulcherrima*	Braid nine branches to be worn around the neck of nursing mother to increase milk supply—also bathe breasts with a tea made of plant before nursing.
Wild Poppy *Chicalote* *Argemone mexicana*	Exudes a white milky sap akin to opium. Useful as a sedative, for insomnia and pain, and to calm coughs.
Wild Yam *Cocolmeca* *Dioscorea* sp.	Chop and boil tuber to drink for rheumatism, arthritis, diabetes, anemia, and fatigue.